Telecommunications History & Policy into the 21st Century

Purpose

The purpose of this book is to provide a broad yet succinct view of U.S. Telecommunications history and policy for college graduate and undergraduate students. It is especially targeted for students taking an introductory course in telecommunications policy. Special attention has been given to make the comprehension acceptable for international students who meet TOEFL standards.

Telecommunications History & Policy into the 21st Century

Ronald G. Fulle

RIT Press
Rochester, NY

Published and distributed by
RIT Press
90 Lomb Memorial Drive
Rochester, New York 14623-5604
http://carypress.rit.edu

Designed by Marnie Soom

All trademarks and illustrations are property of their respective owners and used with permission.

Library of Congress Cataloging-in-Publication Data

Fulle, Ronald G., 1947–
  Telecommunications history & policy into the 21st century / Ronald
G. Fulle.
       p. cm.
  Includes bibliographical references and index.
  ISBN 978-1-933360-39-3 (alk. paper)
  1. Telecommunication—History. 2. Telecommunication policy. I.
Title. II. Title: Telecommunications history and policy into the 21st
century.
  HE7631.F855 2009
  384—dc22

                                                        2009005012

Contents

Question and Pledge

Why are the history and evolution of telecommunications and the forces that have affected the telecommunications environment important?

Anyone who has ever studied history or been interested in genealogy need not ask this question. Actually, if you have taken the time to read the question, then it is likely that you already know the answer. If not, or if you aren't sure, then my pledge to you is that by the time you finish this text, you will know the answer to this question and have a greater appreciation for telecommunications in the United States. You may even be able to use this appreciation and knowledge to extrapolate the future of telecommunications within the U.S. and also perhaps globally. Some may even turn this knowledge, appreciation, and ability into opportunity. While the events of the past have been significant and important, the best is yet to come!

Preface

This book, *Telecommunications History & Policy into the 21st Century*, is the first edition of what may well turn out to be a series of several editions on this fascinating, fast moving, ever changing and challenging subject. The speed with which technology, policy, market forces, and security affect the telecommunications industry and environment will determine just when the next edition is launched. If you have suggestions for the next edition, kindly e-mail the author at: rgfiee@rit.edu.

Chapter 1 is both an introductory chapter relative to telecommunications and policy as well as a chapter that chronologically covers that period of time from the beginning of telecommunications up through 1984.

Chapter 2 is a digression from the chronological progression of this textbook. It covers selected essential Telecommunications Technologies, Notions and Concepts. These foundation topics are necessary in order to understand the essence and ramifications brought about by evolution and change within the industry. Chapter 2 specifically discusses Networks, Numbers, Toll-Free Service, and U.S. Telecommunications Policy and selected Concepts including externality, phatic communications, and several varieties of economic regulation. Knowledge and understanding of these provides the reader with a sound foundation for understanding the rest of the text.

Chapter 3 discusses the breakup of the Bell System a.k.a., Divestiture or Modified Final Judgment. This was announced in early 1982 and became effective on January 1, 1984.

Chapter 4 addresses the postdivestiture period up until the Internet Age.

Chapter 5 discusses the Telecommunications Act of 1996 with special emphasis on Title I services.

Chapter 6 covers the New Millennium through 2009 and includes the Internet Dominance and VoIP, the so-called Triple Play, Convergence, and more.

Chapter 7 is a Crystal Ball into the Future.

Chapter 8 may be considered a Synthesis and Conclusion as well as The End of the Beginning: A New Beginning!

The Appendix discusses three cases via examining specific telecommunications issues and explicit corporations. The companies include: PAETEC, Embarq, Ontario and Trumansburg Telephone Company, and Fibertech Networks. Following the Appendix is a glossary of terms.

I started writing this textbook in November of 2006 and completed it during the fourth quarter of 2009. Fortunately, I was on sabbatical during December of 2008 through early March 2009 and was able to write unimpaired by teaching and other professorial responsibilities during this period of time as well as during two summers. Many people helped me during the course of writing this textbook. I would especially like to thank the following people: Kruti Bhatt, Arunas Chesonis, Dan Clifton, Franz Foltz, Paul Griswold, Robert Kern, Gidon Lissai, Ann Mary Masterson, Debottym Mukherjee, John Purcell, Gregg Sayre, Ganesh Vidyadharan Girija and Laurence Winnie. Also, I wish to thank the entire undergraduate and graduate Telecommunications Policy classes that used a pre-publication manuscript version during their fall of 2008 classes and then provided valuable input to me.

I sincerely wish to thank the folks at RIT's Wallace Library Scholarly Publications center for early research, assistance, and editing including Marianne Buehler, Sherlea Dony, and Nicholas Paulus. Special thanks also go to the mainstays at the RIT Press including editors Molly Cort and Amelia J. Hugill-Fontanel and design specialist Marnie Soom, copyeditor Paul Remington, and RIT Press Director, David Pankow. Thank you!

—Ron Fulle

The Beginning through Predivestiture

The Beginning

Newton's Telecom Dictionary, that so-called *bible of telecommunications*, defines telecommunications as: "the art and science of communicating over a distance by telephone, telegraph, and/or radio. The transmission, reception, and the switching of signals, such as electrical or optical, by wire, fiber, or through the air means."[1] However, in checking the increasingly used but sometimes controversial free online encyclopedia Wikipedia[2], it takes a much broader historical view of the term *telecommunications*. Likely this is due to the etymology of the term: the Greek preface *tele* means far off or at a distance, while the latter part is from the Latin *communicare*, which means share. Wikipedia includes the early North America and China smoke signals as well as drums in South America and Africa. Earlier still were the signal fires in ancient Greece 1300–1100 BCE which were used to flash signals from mountain peak to mountain peak. Homer described this in the *Iliad*. Some of the first messengers on horseback courier systems were implemented during the second millennium BCE in Egypt and the first millennium BCE in China.[3] We all know that a marathon run today is 26.2 miles in distance. The term *marathon* comes from the run of Pheidippides, a Greek soldier who ran from Marathon to Athens (approximately 26.2 miles) to announce the Greek victory over Persia in the Battle of Marathon and to warn Athens that the Persians were headed their way and planned to attack Athens before the Greek army could reassemble there. Legend has it that he ran the entire distance without stopping but that after he announced the victory to the Athenians, he collapsed–dead tired. The circumstances surrounding this communication make it one of the most famous examples of communicating at a distance.

Semaphore or optical telegraph was first outlined conceptually

Demonstration of Claude Chappe's
Semaphore System[5]

to the Royal Society of England in 1684 but not actually deployed until over one hundred years later when Claude Chappe and his brothers succeeded in covering France with 566 tower stations that were used for military and national communications until the 1850s. The French Revolution was a period of great turmoil in France. France was surrounded by the allied forces of England, the Netherlands, Prussia, Austria, and Spain, and the English fleet held Toulon. Researching Chappe's initiatives to provide reliable and swift communications during these difficult times makes for interesting reading.[4] Special circumstances fostered the need for better communications. Once again, focus and motivation brought about innovation.

The point of this discussion on the very early history of telecommunications is that history has proven that humankind has consistently felt the need to communicate both up close and personal and also at a distance. There are many milestones in the history and evolution of telecommunications in the United States. The year 2009 marks the twenty fifth anniversary of the breakup of the Bell System, a.k.a., Divestiture. It is fitting that one step back from all that has transpired in this industry to date and more fully examine it, as the future offers even more change, challenge, and opportunity.

This text provides a valuable examination of the history and evolution of telecommunications within the United States. The approach is not solely based on technological invention, but also includes circumstances, profit, and user needs, which are bundled into market forces. The approach also includes public policy because it has played a very significant and forceful role in helping to shape the industry. Some might say that recent decisions relative to public policy directed toward the telecommunications industry have had a greater effect on it than recent technological advancements.

The last major force that shapes the telecommunications environment is security. During the early years, security in communications and networking was somewhat taken for granted because telephone company facilities were very secure, often had guards at their doors, and their employees were taught to keep customer conversations private. *Listening-in* on conversations was inappropriate and only done when a need to know was paramount. A violation of privacy or security standards was a sufficient cause for dismissal. There was a time when one never had to worry about a virus affecting one's communications device or someone *tricking* distant people or machines into thinking that they were you when, in fact, they were not as in today's "spoofing" schemes. Without adequate security in today's telecommunications networks there would be chaos. Users would not have the feeling of confidence that is necessary to carry on electronic commerce or other advanced communications applications that we take for granted today, such as: working at home, e-mail, and online banking. As humankind's reliance on communications continues to grow, so too will the importance of security. Unfortunately, the public's awareness of this importance may not be fully realized until or unless there is a major security breech. This book will not further discuss or emphasize the necessity of security. Nevertheless, it is important that you the reader understand that security is a major and necessary force in shaping the telecommunications environment in the United States. Together these four forces: technology, market forces, policy, and security make up the so-called "four-legged stool" that supports and determines the direction of the telecommunications environment in the U.S. Innovation, progress, and change are integral parts of each leg of this stool and there may also be

social consequences and outcomes that occur.

If you've had the benefit of taking a business or economics course, perhaps you've heard of other so-called stools that were once in vogue for describing important practices for success. The Bell System considered customers, employees, and shareholders to be the legs of their success stool. There has also been used a concept of marketplace success based upon a so-called value equation.

One such equation was used in the 1990s by the large business sales division of AT&T. It suggested that large business customers should make their network purchasing decisions based upon which provider could offer the best overall value to their bottom line. Purchasing decisions should never be based upon price alone but should be based upon overall value. Value should be the sum of the following: quality, reliability, service, innovation, and price. The equation would be: $V = Q + R + S + I + P$. In each of these examples, the issue is really the individual success or failure of one company. Do not confuse the success or ups and downs of individual corporations with the overall telecommunications environment depicted by the four-legged stool.

Early Years 1838–1894

Many consider so-called "modern" telecommunications to have been founded with the introduction of the first all-digital network–the telegraph–in 1838, when the U.S. Congress gave Samuel Morse $30,000 to build facilities between Washington, D.C. and Baltimore, Maryland. The first public demonstration of this service occurred on May 24, 1844, when Samuel Morse sent the message, "What Hath God Wrought?" between these two cities.[6] While these facilities were publicly funded, Morse privately held his patents. In 1846 Morse offered his patent to the federal government for $100,000, but Congress declined to accept the offer. This action provided a tremendous impact on the United States telecommunications environment compared to telecommunications in nearly all other countries. The difference is that in the United States, telecommunications providers grew up in the private sector rather than the public sector. Only for a brief period of time during the latter part of World War I did the U.S. federal government take over the telephone company (and the railroads) for national security and defense reasons. On August 1, 1918 the U.S. Postmaster General assumed supervision, possession, control and operation of telephone properties and returned this control to the Bell System on July 30, 1919.[7]

A more traditional beginning of "modern" telecommunications is thought to be the invention of the telephone. In 1874 Alexander Graham Bell decided that it was possible to put speech over wire by varying the electrical current, as the densities of air waves vary when sound is produced. On April 6, 1875, the U.S. Patent Office issued Patent No. 161,739 to Bell for the telautograph. However, the patenting was suspended pending an interference which means that another inventor's claims conflicted with his. This other inventor was Elisha Gray. Bell's lab work continued. On February 12, 1876, Washington lawyers again filed Bell's patent for the telephone. Later that same morning, Elisha Gray filed a patent caveat for a telephone, although he had not yet built one and Bell had. In addition to Gray, Paul La Cour of Copenhagen and Thomas Edison also filed for patents. On February 29, 1876, Bell went to Washington and was eventually able to show that his earlier filings included everything that Mr. Gray had in his caveat and also superseded all other claims. Bell went back to Boston and resumed his experiments and later was advised that his patent was eminent. Alexander Graham Bell's telephone patent number 174,465 for "Improvements in telegraphy" was issued on March 7, 1876.

This patent turned out to be one of the most challenged and valuable patents ever issued. On March 10, 1876, the first complete sentence, "Mr. Watson, come here, I want you!" was accomplished via this new telephone technology. While Bell and Watson were experimenting, Bell in one room with the transmitter and Watson in another with the receiver, Bell spilled some acid from the transmitter onto himself and then called Watson for help.[8] It is interesting that the discovery and patenting of the telephone by Alexander G. Bell in 1876 simultaneously marked the birth of the technology and the birth of the telephone business. Bell was a brilliant inventor and an extreme *tinkerer*. Bell soon lost interest in his telephone invention and sold many of his patent rights. He used the considerable proceeds from the sale to pursue his passion of tinkering with airfoils, hydroplanes and a variety of other techno-toys of the day. His financial backers recognized the potential value of the invention and, with Bell's expert testimony, successfully fought more than 600 court battles over Bell's patent rights. The American Bell Telephone Company was formed in 1877. It grew and prospered via the exploitation of the proprietary patents until their expiration in 1894. This was in keeping with the patent convention that still applies today, allowing inventors who hold patent rights to have exclusive use of the invention for a limited period of time in order to set prices above costs and reap their just rewards. Upon the expiration of the patent, the technology becomes available to all, and this usually results in competition and its resulting consumer benefits. For the original patent holder though, the period immediately after the time when the patent expires is thought to be one in which "all hell breaks loose."

Competitive Years 1894–1907

This is exactly what happened: when the Bell patents ran out, hundreds of competitors entered existing markets and opened up new markets, driving down the price of telephone exchange service and capturing market share. By 1902, independent telephone companies served 44% of the market, and by 1907 their market share had increased to 51%.[9]

American Bell Telephone was losing market share and its return on investment was spiraling downward. Its financial position deteriorated so badly that in 1907 it fell into the hands of the bankers led by J. P. Morgan, whose syndicate owned $90 million in unsold AT&T bonds. Morgan fired the entire board of directors and created American Telephone and Telegraph now simply named AT&T the successor holding company to American Bell. He also hired Theodore N. Vail to lead the new AT&T. Vail had worked for Bell and Western Union earlier in his career.

Decline in Competition 1908–1918

When Vail took over, AT&T was a mess and so, too, was the entire industry. There was fierce competition, poor service, a terrible public image, and no interconnection of the independent telephone companies.[10] Subscribers to telephone company "A" could not call subscribers to telephone company "B." This system of dual and independent services was a byproduct of competition resulting from the expiration of the Bell patents in 1894 and the noncooperative approach that Bell and these independent (a.k.a., non-Bell) telephone companies pursued. A similar situation took place in the 1990s when instant messaging leader AOL refused to cooperate with other IM providers. They too seemed to desire to profit and grow by not cooperating with their competitors and by ignoring the good of the general public.

Vail had tremendous effects on AT&T and the evolution of the telephone industry in the United States. In 1908 the term "Bell System" was introduced via national advertising. Also, Vail originated the theme, "One Policy, One System, Universal Service."[11] Universal service was Vail's concept of providing residential telephone service priced sufficiently low so that anyone in the United States could afford it–even if they were miles from town. This concept not only helped the consumer but caused other telephone services to be priced slightly higher to help defray these universal service expenses. In effect, a process and system of subsidy was created to help defray the high cost of telephone service for those who were a long distance from town. This also helped to grow the system of telephones, which not only helped those living in rural areas but also greatly helped the Bell System. While Theodore

Vail was spreading philosophies and telephones as far and wide as he could, he also recognized that the only way to save the company was to develop proprietary technology. He hired two physicists named Pupin and Campbell. In 1900 these physicists developed and *patented* the "loading coil".[12] This device provided the next major technological advantage for AT&T in that it provided the ability to carry long-distance calls. Prior to its invention, short-haul traffic of less than ten miles was the norm. While AT&T was now able to carry long- distance toll calls, it continued to refuse to interconnect with other telephone companies. By 1913, customers were flocking to AT&T and its operating companies due to its ability to carry long distance calls, and Vail was purchasing financially troubled independent telephone companies at a fraction of their worth and thereby expanding the business. To say that a company is an independent telephone company means that it was never part of the Bell System. This scenario reminds me of that classic and famous arcade game, *Pac-Man*, with Bell being *Pac-Man* and all other "telcos" being one of the colored ghosts that, if unsuccessful, are gobbled up by *Pac-Man*. While *Pac-Man* was a fun game to play, few of us likely envisioned ourselves as the preyed upon in that game. Yet there were many independent, a.k.a. non-Bell, companies that likely experienced that feeling.

An interesting struggle occurred in Upstate New York. Around 1905, a handful of prominent residents from Rochester, N.Y. and several business leaders from across the country formed the United States Independent Telephone Company. Its intent was to meld together the independent telephone companies in a widespread network that could challenge the dominance of the Bell System. Among the twenty-six directors were George Eastman, Edward Bausch, Hiram Sibley, and Henry Strong. This entity became a holding company and acquired the controlling interests of the Rochester Telephone Company; the Kinloch Telephone Company of St. Louis, Missouri; the Home Telephone Company of Kansas City, Kansas; the Federal Telephone & Telegraph Company of Buffalo, and others. It bought Stromberg Carlson Manufacturing Company of Chicago and brought it to Rochester to be its equipment manufacturing arm, much like Western Electric was to the Bell System. Its success hinged on the granting of a franchise to the Independent Telephone Company of New York City. This was expected to be routine. However, the franchise was not granted. Without the populous New York City entity, its stock plummeted from $45 per share in 1906 to $3 per share fourteen months later. The company was dissolved in 1907 and many investors lost money.[13]

As a direct result of Bell buying up its competitors, in 1913 the U.S. Department of Justice (DOJ) filed an antitrust complaint against AT&T. The result of this action was a late 1913 out-of-court settlement called the Kingsbury Commitment (due to it being handled by an AT&T Vice President by the name of Kingsbury). The major provisions were:

1. AT&T agreed to divest itself of Western Union — the once-mighty telegraph company. It acquired control of 30% of Western Union by stock purchase in 1910 and Theodore N. Vail was president of both AT&T and Western Union from November 1910 until April 1914.

2. AT&T agreed to buy no additional independent telephone companies without approval from the Interstate Commerce Commission (ICC). (The ICC regulated telephony prior to the creation of the FCC in 1934.)

3. AT&T agreed to *interconnect* with other telephone companies.

This so-called Kingsbury Commitment was a political solution to the problem of Bell's monopoly and power versus the right of independent telephone companies to exist and to prosper. It was an important policy direction change as it basically established a partnership between Bell and its competitors. Customers served by independent telephone companies desiring to call long distance could now dial long distance and the independent Telco would pass the call through to Bell, who would complete the call. Revenues were shared for these calls by both Bell and the independent telephone companies. This would come to be called the division of revenues settlement process. In addition, wherever there were competing Bell and non-Bell telephone companies, such as in Upstate New York, it soon became obvious that it would be more economical and efficient if one company provided the telephone service within a geographical boundary. Rochester; Jamestown; and Buffalo, New York all had dual competing telephone companies. In 1915 a principle to consolidate was negotiated. Bell was to acquire the independent companies' assets in Buffalo, while Bell would give up its assets to the independent companies in both Rochester and Jamestown. This was an approximate equal swap. Many details needed to be worked out and duplicate telephone service needed to be eliminated. The consolidation became official in 1921 and ended the two-telephone era in Upstate New York.[14]

Theodore Vail recognized the tremendous advantage and protection that research and development provided the Bell System, such as the telephone and loading coil patents. Consequently, he is credited with establishing the Bell tradition of scientific research that, for many years, kept it in the vanguard of technological innovation. This tradition helped form the Bell Telephone Laboratories, Inc. in 1925, when the Western Electric Research Laboratories and part of the AT&T Engineering Department were consolidated. Bell Laboratories was *the* preeminent U.S. research facility for the better part of fifty years.

Another great visionary initiative that Vail also is credited with is advocating the philosophy that telephony was a *natural* monopoly, which meant that there should be only one provider of telephone service for a given geographic area and that this provider would necessarily come under some kind of governmental regulation. Vail said, "a regulated monopoly best serves the public."[15] Additionally, public services such as telephone service were best served by a single source. This concept was coupled with the concept that maximum private profit was not the primary objective of telephone companies. Profit was only necessary to maintain the health of the corporation. These philosophies undoubtedly helped bring about the Kingsbury Commitment, which set the tone for the telephone industry in the United States for the next seventy years. Basically, the public and government accepted Vail's concepts of universal service and "natural monopoly" with government oversight and cooperation.

Height of AT&T's Power 1918–1949

In 1934 Congress passed the Communications Act, which was the foundation for federal government regulation of the industry. The Communications Act of 1934 created the Federal Communications Commission (FCC) and designated it as being responsible for telecommunications regulation, including the enforcement of the Kingsbury Commitment. It also confirmed that the telephone industry operated best in the public interest as a regulated monopoly. Immediately after its creation, the FCC began an indepth and ongoing study and investigation of AT&T. Meanwhile, AT&T continued to grow in size and power and significantly prospered during these years. In 1929 there were 500,000 stockholders in the Bell System and it became the first company to achieve $1 billion in revenue. By 1939 AT&T had assets of $5 billion, served 83% of the telephones, owned 91% of the telephone plant, and provided 98% of the toll service and 100% of the international or transoceanic telephone traffic. There were also 600,000 stockholders who worked to Bell's advantage by creating a large and favorable base for public opinion and it also helped to ensure that no small group could control the company. AT&T weathered the Great Depression years in fairly good shape. It never failed to pay the dividend, and, rather than lay off workers, AT&T reduced its hours and kept all workers employed. These two actions yielded fiercely loyal shareholders and employees. As the U.S. economy improved and World War II came and went, AT&T continued to grow and prosper.

After many positive and some negative actions directed toward the Bell System, in 1949 the Justice Department filed a new suit under the Sherman Antitrust Act in an attempt to introduce competition into the manufacturing and supply of telephone equipment. The investigation that the FCC began soon after it was created contributed to this action. The DOJ charged that Western Electric's (WECo) products were subsidized by telephone rate payers, giving WECo an unfair advantage over its competitors. The following is a summary of the charges of the suit:

1. Effective (equipment) competition was absent.
2. Because of the vertical integration of AT&T and Western Electric Co., WECo could overcharge AT&T and this would be addressed by the AT&T Operating Companies charging end users via the rate-making process.
3. WECo should be divested from the rest of AT&T.
4. AT&T should be split into three companies.

Competition and Deregulation 1949–Predivestiture

While this suit was progressing, AT&T continued to expand its size, scope of operations, and number of shareholders. Bell Laboratories was recognized as a national resource and the government even requested AT&T to take over what was to become Sandia Labs, a research and scientific weapons lab in Albuquerque, New Mexico. The Justice Department was attempting to split up AT&T, while at the same time other branches of the government both treasured and sought it to become a good corporate citizen and to provide help vital to the security of the United States. The 1949 lawsuit dragged on and was eventually settled out of court by the 1956 Consent Decree, which established that:

1. WECo was not separated.
2. WECo could not sell their products outside the Bell System (except to the government).
3. AT&T would engage in common carrier business only.
4. AT&T agreed to grant nonexclusive license and technical information to any applicant, thereby eliminating most patent benefits.

At the time many people felt that AT&T had won a very major victory in this settlement. AT&T wasn't split up. Western Electric was not divested. However, a closer look has shown that by providing free access to the technological advances of Bell Labs, all technology breakthroughs resulting from their research and development could no longer be used to a proprietary marketing advantage during the patent rights protection period. This included the basic transistor patents, laser, and all cross-licensing agreements of 1926.

Theodore Vail had instituted two strategic foundation principles that had endured for over fifty years for the Bell System. They were: technological dominance and service provided via a regulated, single source, and protected monopoly. The concession to give free access to Bell Lab's technological advances fostered the elimination of technical dominance. In fact, the greater and more significant a technical breakthrough, the faster a new competitor (perhaps one not encumbered by tremendous existing invested capital and legacy products and systems) could apply the new technology and thereby steal customers from the

Bell System. That is to say, a potential competitor might steal customers from the Bell System if not for Bell's strong and significant protection as a regulated monopoly. However, this protection was being challenged.

The first chink or flaw in the regulated monopoly policy armor of the Bell System was a small and relatively insignificant suit called the Hush-a-Phone case, which involved a company that manufactured a plastic device to put over the mouthpiece of a phone to keep other people from hearing background noise. It worked much like cupping one's hand over the mouthpiece. Initially, Hush-a-Phone lost the case, but eventually it did win based on the fact that AT&T could not prove that the device could harm the network. There was little demand for this device, but its significance is that it was the first chink in the armor. Of greater significance was the Carterfone case. In June of 1968 the FCC struck down existing interstate telephone tariffs prohibiting attachment or connection to the public telephone system of any equipment or device that was not supplied by the telephone companies. This was another in a very long series of lawsuits that AT&T lost. The result of these judgments, coupled with the October 18, 1977 FCC-initiated registration program, was that equipment from providers other than the Bell System could be directly connected to the Public Switched Telephone Network (PSTN). Today, for the wired marketplace at least, you can readily purchase telephone sets, answering machines, and other telephone instruments from many, many outlets, and they are produced by a wide array of manufacturers. Each offers its own price, value, and feature levels. As a result of the Carterfone decision and ensuing competition, coupled with a distinct lack of highly competitive telephone terminal equipment, AT&T lost a substantial share of the U.S. telephone terminal equipment market in the late 1970s and 1980s. Having given up reliance upon regulated monopoly protection in this market, AT&T began focusing on business and residence telephone service and business telephone terminal equipment. It also desired entry into the perceived highly lucrative computer marketplace.

While the telephone terminal equipment market was opening up, long distance telephone service, which was the heart of the Bell System, was simultaneously under attack via a series of lawsuits. As early as 1963, a young upstart company named Microwave Communications, Inc.

(MCI) attempted to capitalize on the *Above 890* decision, which was an FCC ruling that allowed private firms to own and operate, for their own use, private microwave systems at frequencies above Bell's standard frequencies. MCI wanted the *Above 890* decision to be modified so that it could resell private microwave service between Chicago and St. Louis. AT&T charged that this would effectively be cream-skimming and, in the long run, could adversely affect universal service. The FCC chose to not address these larger issues and only focus on MCI's application for the Chicago-to-St.-Louis service to which it granted final authority and approval in 1971.

"During the time period of 1974 to 1981, the rate of regulatory change increased a hundredfold over the previous decade."[16] In almost every case, AT&T fought hard to retain the status quo of a regulated monopoly, protected from competitive local and long distance service providers by the basic laws of the land, while would-be competitors fought equally hard to gain entry into Bell's markets. In most cases, pro-competitive decisions were handed down by the FCC. In 1974, the U.S. Department of Justice (DOJ) filed a most significant antitrust case against AT&T alleging monopolistic practices and calling for its breakup. This was addressed in an out of court agreement and the announcement came on January 8, 1982 when the U.S. DOJ and AT&T announced the breakup of the Bell System. This became known as the Modified Final Judgment (MFJ), or merely divestiture, as AT&T divested itself of the twenty-three operating companies.

Summary

The history and evolution of telecommunications from the very beginning of communicating at a distance through 1983 is strongly influenced by technology, public policy, and market-driven factors including the desires and needs of individuals and the public. In the beginning there was a need to communicate at a distance. When Alexander Graham Bell invented this marvelous wonder in 1876 to address this need, he also coincidently launched the telephone industry. This industry went from the early years of little or no competition due to patent rights on the telephone to intense competition and chaos soon after those patents expired. There was then a period of over fifty years of a regulated monopoly and the development of the finest telephone system in the world. It was a system in which technological and market leadership, as well as much of the basic service, was provided by one large corporation called AT&T and often referred to as the Bell System. The Bell System interconnected with the many large and small independent non-Bell telephone companies and they all worked together to provide excellent nationwide telephone service and arguably the best telecommunications service in the world.

Interestingly, approximately midway though the twentieth century, and with little regard to having the best telephone service in the world, the U.S. government and many individuals and corporations began to challenge the concept of a single-source provider, as well as the concepts of protected markets and the resulting monopoly this provided the Bell System even though the Bell System clung to its natural monopoly concept. The challenges surrounding the U.S. telecom environment played out in the various public policy venues including those of: U.S. patent policy, DOJ suits and settlements, Congress via the Communications Act of 1934, and the FCC. Public opinion and public policy moved from that point in time when the provision of telephone service via a regulated monopoly was thought to be good to a point in time when the competitive model became king. The breakup of the Bell System was intended to help usher in this new competitive model. Chapter 3 of this book is dedicated to the breakup of the Bell System, a.k.a. the Modified Final Judgment (MFJ) or just plain divestiture.

Alexander Graham Bell

Alexander Graham Bell was a remarkable and imaginative man. He was not a super manager or industrial giant in the mold of Andrew Carnegie or Nelson Rockefeller. Nor did he demonstrate tremendous ability to move into the ranks of the super wealthy. Yet he was remarkable and a man of recurring paradox. "Here was a man who made a stunning technological advance not only *in spite of* but also *because of* his lack of training in the field. He came to his miracle of sound transmission in working to help those who would be totally unable to avail themselves of it, including his own deaf wife and mother."[19] He was clumsy too, but this very characteristic, coupled with the lab expertise of his assistant Thomas Watson, helped him to invent the telephone.

It was said that children loved Bell. "His marriage was a model of devotion throughout its forty-five years. He was lionized in society, cheered at exhibitions, applauded at scientific meetings, and sought out by reporters."[20] He was a kind and decent man and people liked Alexander Graham Bell.

In addition to inventing the telephone and writing such a fine patent that it withstood over 600 lawsuits, he spawned many other inventions. He improved Edison's phonograph and devised the metal detector, the hydrofoil, and the respirator. He and his associates conducted over 1,200 experiments in aviation and achieved the first public airplane flight in the United States. The Wright Brothers flight was privately held. They feared that their machines would be copied. Wilber and Orville Wright "refused to fly in public until the American or French governments agreed to buy their airplanes on acceptable financial terms." Bell helped to found the National Geographic Society and its magazine. He was instrumental in bringing Montessori education to America and was an early civil rights advocate. In 1914 Bell coined the phrase "greenhouse effect" in an essay on the threat of global warming. While a top inventor by any standard, Bell himself viewed his work in the education of the deaf as his life's work. He had a lifelong campaign to teach deaf people to lip read and to communicate.

From his boyhood he had tended to be aloof and solitary. He himself wrote in 1894, "I somehow or other appear to be more interested in things than people..." He recognized "the tendency to retire into myself and be alone with my thoughts." Yet he struggled against this with the help of his wife. Ironically, that lifelong struggle may explain in part the intensity of his dedication to bring "the human family in closer touch." Upon Bell's death on August 2, 1922, his archrival and later friend Thomas Edison praised him for having "brought the human family closer in touch." Though

Edison had only the telephone in mind, the thought and thread seemed to characterize Bell's whole range of achievements.

Alexander Graham Bell was born in Edinburgh, Scotland, and retired to Baddeck Bay, Nova Scotia, Canada where he built the family home "Beinn Bhreagh" or "Beautiful Mountain." He died at the age of seventy-five and is buried there. Upon his death, "employees of the American Telephone and Telegraph Company sent a large wreath of laurel, wheat, and roses. His wife Mabel explained it to her daughter: 'laurel for victory, wheat for the gathered harvest, and the roses for gentleness and sweetness.'"[21] His epitaph reads, "Died a citizen of the United States."

Alexander Graham Bell Speaking into a prototype telephone 1876[17]

Portrait of Alexander G. Bell c. 1910[18]

All information in this "Bell Box" comes from the excellent, fully-illustrated and highly recommended book: *Alexander Graham Bell the Life and Times of the Man Who Invented the Telephone* by his great grandson Edwin S. Grosvenor and Morgan Wesson with a forward by technology historian Robert V. Bruce, who earlier wrote: *Bell: Alexander Graham Bell and the Conquest of Solitude*.

Chapter I Study Questions

1. Research and explain the Hush-a-Phone case.

2. Research and explain the Carterfone incident in telecommunications history.

3. Research and explain one additional interesting fact about Alexander Graham Bell.

4. When MCI sought permission to offer microwave service between Chicago and St. Louis, AT&T charged that this would effectively be cream skimming. Why?

5. What part, if any, did radio play in the evolution of telecommunications?

6. Search/Google "Milestones in AT&T history" and learn more about other various milestones from the AT&T hosted milestone site. From this site you should be able to determine the answers to the following:
    a. When was Bell Laboratories established?
    b. When was the first transatlantic telephone service established?
    c. When did Bell Laboratories win the first of its many Nobel Prizes?
    d. When did AT&T first offer mobile telephone service?
    e. When did AT&T introduce the first commercial modem?
    f. When was touchtone service introduced?
    g. When did AT&T introduce 911 as a nationwide emergency number?
    h. When did AT&T develop UNIX?
    i. When did scientists at Bell Labs first invent the transistor?
    j. When did AT&T install the first commercial fiber optic cable in a commercial communications system?
    k. When did the Bell System cease to exist due to divestiture?
    l. When did AT&T announce their restructuring into three separate companies, a.k.a. "trivestiture"?

Notes to Chapter 1

1   Harry Newton, *Newton's Telecom Dictionary*, 22nd ed. (Berkeley: CMP Books, 2006).

2   Wikipedia, "Telecommunications," Wikipedia, http://en.wikipedia.org/wiki/Telecommunications.

3   Anne B. Keating and Joseph Hargitai, "Ancient Networks," (New York: New York University Press), http://www.nyupress.org/professor/webinteaching/history3.shtml.

4   Wikipedia, "Semaphore," Wikipedia, http://en.wikipedia.org/wiki/Semaphore.

5   "Semaphore Line Redirected from Optical Telegraph," Wikipedia, http://en.wikipedia.org/wiki/Optical_telegraph.

6   AT&T Archive, *Events in Telecommunications History*. 1.

7   Ibid., 38.

8   Ibid., 5–6.

9   Ibid., 255.

10 Ibid., 255.

11 Ibid., 30.

12 Ibid., 26.

13 F. L. Howe, *Endless Voices—the Story of Rochester Tel*, (1992), 18-20.

14 Ibid., 22–23.

15 AT&T Archive, 255.

16 Gerald R. Faulhaber, *Telecommunications in Turmoil*, (Ballinger Publishing Co., 1987).

17 "Alexander G. Bell," Wikipedia, http://en.wikipedia.org/wiki/Alexander_Graham_Bell.

18 Ibid.

19 Edwin S. Grosvenor and Morgan Wesson, *Alexander Graham Bell: The Life and Times of the Man Who Invented the Telephone* (New York: Harry Abrams, 1997), 6-7.

20 Ibid., 7.

21 Ibid., 289.

CHAPTER 2

A Digression: Basic Telecommunication Technology, Policy, and Concepts — Then, Now, and in the Future

①There are certain telecommunications technologies, policies, and concepts that are essential to know in order to truly understand the essence and ramifications brought about by evolution and change within the industry. The intent of this chapter is to help you toward this enlightenment relative to these topics, thereby providing you with not only a sound foundation of understanding but also a basis for interpretation and projection forward. This chapter will explicitly discuss Networks,② Numbers, Toll-Free Service, U.S. Telecommunications Policy, and selected Concepts with just enough depth to provide this foundation. Readers may wish to further research some of these topics on their own, as there is ample information available on each of these individual topics to compose volumes. Additionally, there are end-of-chapter suggested research questions that, when answered, will yield further learning and discussion opportunities on key topics.

Networks

Networks are of many and varied types. There are human networks consisting of business, entrepreneurial, social, economic, *good old boy*, and dating, to name a few. There are media networks that might include radio, TV, and gaming. Technical networks would include computer and ③ telephone for sure and may be wired or wireless and have a local (LAN), campus (CAN), metropolitan (MAN) or wide-area network (WAN) scope. Often these are studied via breaking them into smaller "bite size" pieces using either the OSI or TCP/IP reference model criteria. Technical networks might also include energy transmission networks like electrical, gas, and oil. Within the fields of mathematics, science, and engineering, there are flow networks and neural networks, and network theory or *diktyology* is a subject within applied mathematics and physics that coincides with graph theory. There are likely many other unique and specialized networks, a few of which are: religious networks, spatial networks, storage networks, and transportation networks. Beyond these, some may even believe in and study *paranormal* networks.

So as to not exhaust you and to help us maintain focus on topic, this section will address technical networks from the vantage point of computer-oriented telephone networks. Suffice it to say that twenty-five years after divestiture, many of the networks listed use these types of technical networks and, in some cases, are converging into hybrid communication networks. For instance, today most media networks run on these computer-oriented telephone networks and much social networking is done through them as well. Modern carrier networks are converging voice and data networks into hybrid IP networks and smart business entities are taking advantage of this convergence opportunity. Today's networks are benefiting from the great breakthroughs in fiber optic transmission and other technological advances. However, the most significant change agent in the field of networking today is the movement from so-called *traditional* circuit-switched networks to IP networks. The underlying reason is simple—economics. Properly featured and configured IP networks offer a great financial advantage over traditional circuit-switched networks for everyday voice calling. While a great advantage, some traditionalists view this phenomenon as destructive to the set of current computer and telephone networks and the laws, revenue streams, policies, and structures that arose during the heyday of the current PSTN. While destructive, it doesn't really matter, as change often butts up against restraint. When properly founded, change will occur.

Let's first look at traditional circuit-switched computer and telephone networks which include the Public Switched Telephone Network or PSTN that most of us used for voice calling during the twentieth and into the twenty-first century. This is sometimes referred to as the POTS or *plain old telephone service* network. The trend toward today's modern PSTN architecture began with the introduction of the transistor, computer, and stored-program-controlled switching systems, which were introduced in the mid-1960s. Intelligence and efficiency were provided to the network by the marriage of computer and telecommunications technology and was manifest by advanced features of the day such as three-way calling, speed dialing and call waiting.

(10)   Telecommunications networks became more intelligent in the late 1970s when the first common-channel signaling (CCS) links were installed for routing call-setup and other call-control information between central office switches.[22] CCS links are special, separate from the call path links or paths and carry signaling information regarding the call. It is, therefore, referred to as out-of-channel signaling. Signaling provides three basic functions that nearly all traditional or legacy telephone systems require:

- Supervision is monitoring the status of a circuit to determine if it is busy, idle, or requesting service.
- Alerting indicates the arrival of a call and is most commonly recognized when a phone rings.
- Addressing is the transmission of routing and destination signals.

(11) Prior to the introduction of CCS, signaling information commonly had been carried on the same circuit that carried the call and was referred to as in-band signaling. This type of signaling requires that all of the link parts between the origination and destination points be actually established end to end. One can readily understand how wasteful this would be, when so many calls do not complete due to a busy signal or no answer. With this in mind, the out-of-channel signaling was truly a major breakthrough. Today's out-of-channel signaling is commonly re- (12) ferred to as Common Channel Interoffice Signaling or CCIS. CCITT Signaling System # 7 (SS7) is a specific version of CCIS and generally is deployed in today's circuit-switched networks.

In 1984 Bellcore began the development of a family of software (13) systems to support 800 number services by building upon SS7. Bell Communications Research, or Bellcore, was a consortium originally established by the seven Regional Bell Operating Companies (RBOCs) and created from Bell Labs in 1984 at divestiture. It was originally owned by these seven RBOCs and was that portion of Bell Labs dedicated to the research and development of local exchange telecommunications technology. "In 1997 the company was acquired by Science Applications International Corporation (SAIC). Since it no longer had any ownership connection with the Bell regional companies, the name was changed to Telcordia. The company was subsequently sold in November 2004 to Providence Equity Partners and Warburg Pincus, who currently both hold equal stakes in the company."[23] The original post-divestiture work was part of a larger industry-directed effort toward creating a network architecture that would make the introduction of new services more efficient. The resulting network architecture that was put into place in the late 1980s and early 1990s came to be called the Intelligent Network (IN).[24]

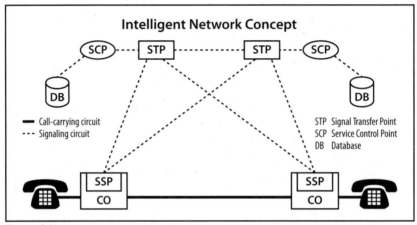

Figure 2.1

The IN utilizes many advances in hardware, software and, philosophy of call processing. Let's review some of these and understand how a call flows. First, note the solid call-carrying circuit. The actual call travels over this circuit. However, the signaling circuitry is the SS7 interoffice signaling circuit represented by the dashed lines. A service switching point (SSP) is a switching system software product that allows for the recognition of calls requiring special handling, like an 800 call. The SSP typically does not have the instructions for processing these special calls–only the intelligence to recognize them. The SSP obtains the necessary instructions for processing the special call by communication with a service control point (SCP) after transferring through a service transfer point (STP). By concentrating more of the software and network logic for services into fewer but intelligent network nodes, services are more easily able to be created and modified without the need for major programming changes in all switches in the entire network.

The Advanced Intelligent Network (AIN) is even more flexible, efficient, and intelligent. It evolved from the IN and is still evolving. With evolutionary changes to extremely large and complex things, it is difficult to say just when the IN became an AIN. Different telephone companies began introducing it at different times. Generally, one can count on the early 1990s as the introduction and movement to AIN. AIN augments IN capabilities by introducing the Integrated Service Control Point, Intelligent Services Peripherals, and the SPACE service-creation system.[25] The Integrated Service Control Point (ISCP) is really the heart of AIN. It is a family of systems for service creation, service management and service execution. Connected to it was Bellcore's SPACE service-creation environment, which supports the design of new services and helps generate the required software functionality. It has a built-in expert system to facilitate and insure feature compatibility and functionality in new services.

One example of the use of SPACE is for single seven-digit telephone number service that can be used throughout the given market area–possibly spanning areas served by several carriers. U S West used this for a large chain of pizza restaurants. A typical call flow to order a pizza would be as follows:

> An incoming call is routed to a SSP where a query is launched to an SCP, which contains the service logic describing how the call should be handled. The SCP queries a Line Information Data Base (LIDB) to determine the location of the calling party. The carrier's Line Information Data Base determines the nine-digit postal code, or "ZIP code + 4" associated with the telephone number of the calling party, and communicates it to the SCP, which translates the postal code into the telephone number of the nearest restaurant. That number is then returned to the SSP so that the call can be completed. All of this takes a fraction of a second.

Figure 2.2 depicts an AIN based upon early-to-mid 90s Bellcore Exchange articles. The issue here is not necessarily the detail of the original or evolved AIN, but rather the fact that all this yielded a highly intelligent call-processing network, and that is significant!

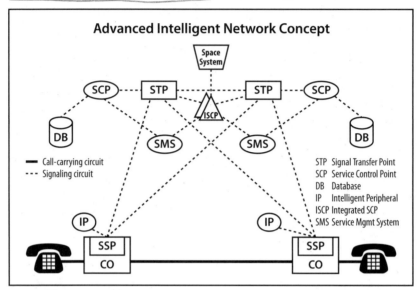

Figure 2.2

Circuit-switched networks were/are used for both voice and data transmissions. As you can see, much detailed and exacting work was done to evolve the circuit-switched network into a very intelligent, centralized, controlled network. The PSTN is a Time Division Multiplexed (TDM) network that requires 64Kbit/sec per typical voice channel. While nice and neat and something that many came to think of as standard, a drawback of this concept is that every voice channel typically uses 64Kbit/sec of bandwidth once the circuit is established, regardless of whether the calling or called party is actually talking. This wastes valuable bandwidth resources and basically is the cause for the demise of typical circuit-switched networks in favor of a better way.

Let us now turn to that "better way" network: the Internet Protocol or IP Networks. IP networking basically grew up as a result of the U.S. Department of Defense's charge to create a highly robust network that would be fast and resilient. The first node of this network went live in the fall of 1969 at UCLA and was eventually named ARPANET. By the early 1980s this network had grown significantly and TCP/IP became operational on it in 1983, when the National Science Foundation (NSF) constructed a university backbone that eventually became NSF-Net. This was opened up to commercial interests in 1985 and the first web pages were born in the very early 1990s.[26] Early adapters first used a web browser named Mosaic in the early 1990s, followed by Netscape, and then followed by Internet Explorer, Safari, Firefox, and others. IP networks are young compared to circuit-switched networks. Unlike cir- cuit-switched networks that use centralized computerized control with TDM and set up 64Kbit/sec circuits, IP networks use the IP protocol to route packets without setting up specific circuit routes through which all following parts of that message would flow. What this allows is an IP network to *not* require a fixed bandwidth for an application that may not be using bandwidth every moment of every second. In an IP network, there is not a time when an application is in need of bandwidth and yet bandwidth is idle. It better matches resources to need and thereby pro- vides for a network that is more efficient than the PSTN, and, therefore, it is more cost-effective for both the provider and user. If modern techniques of Quality of Service (QoS) and security are deployed in the IP network, it can yield call quality and reliability that rivals the excellence that we've come to expect from the PSTN. However, if one merely uses the public Internet without QoS, calling will downshift to a "best effort" status and may sound a bit choppy and generally not work as well. That is why one needs a properly designed and implemented IP network for real-time applications like voice, streaming video, and more. Today Multi-Protocol Label Switching (MPLS) advances basic IP routing by allowing for QoS and what some refer to as "express lane" transmission. Also, it allows for Virtual Private Networks (VPNs) to more easily be created in order to provide an excellent IP network that allows enterprises to cost save via VoIP and merge data and voice networks into one properly designed IP network.

The point of the discussion of these two basic types of networks is not to make you a networking guru. Rather, it is that creative technological advancements yield change that has brought us to a point in time when the creative (albeit destructive to old existing circuit-switched networks and to all that giving up the old in favor of the new entails) IP networks of today and the future are yielding a tremendous paradigm shift. The shift is to a new method of networking for all applications, including voice calling, which at one time was a tremendously large revenue stream, but which now can be done so efficiently so as to significantly reduce the cost of voice calling and, therefore, the revenue stream for providers. One must sometimes wonder if anyone or any provider can make a profit on basic residential long-distance service. Yet that concern diminishes when one looks beyond voice calling, beyond yesterday to future great applications that may replace the voice revenue stream such as applications like HD IPTV and gaming and very high speed Internet with voice calling tossed into the bundle for free. Many consider VoIP to be revolutionary. However, history will likely show that in addition to making voice calling inexpensive, its greatest contribution to the evolution of networking is that it made people take notice of IP networking. This notice helped foster convergence and the many advanced features and applications that are just now coming to market. However, this is a topic for future discussion, so we will leave it for now other than to say that as we pass the divestiture + twenty-five year milestone, the one sure thing relative to networks is that they will continue to change and evolve.

Numbers

Early in the history of telephony, there were no telephone numbers. Individuals called individuals and were assisted by telephone operators. In late 1879 and early 1880, the first use of telephone numbers occurred in Lowell, Massachusetts. According to the AT&T archives, it is well substantiated that during a measles epidemic, "Dr. Moses Greely Parker feared that Lowell's four operators might succumb and bring about paralysis of telephone service. He recommended the use of numbers for calling Lowell's more than 200 telephone subscribers so that substitute operators might be more easily trained in the event of such an emergency."[27] It was feared that the public would take the assignment of numbers as an indignity, but telephone users immediately recognized the practical value of the change. The use of numbers quickly spread.

In 1947, AT&T saw the need for a logical standard for numbering in the United States. This was the post-war era, the population was booming and so was the growth in telephone service. AT&T created a plan that would allow the country to be divided into 152 numbering plan areas which are commonly referred to as NPAs.[28] This plan was the beginning of the North American Numbering Plan (NANP), which now encompasses what is referred to as World Zone I consisting of the entire United States, most territories, Canada, Bermuda and most of the Caribbean.[29] There are now nineteen North American countries or territories in total covered by the NANP and the plan conforms to International Telecommunications Union (ITU) Recommendation E.164, the international standard for telephone numbering plans. The ITU-assigned country code "1" and an international call prefix of "011" have been designated for the NANP area.

AT&T designed, planned, and administered the NANP, largely without intervention until divestiture, when responsibility for the NANP passed to Bell Communications Research (Bellcore). Bellcore administered the plan for a time, but others, including nonwireline cellular carriers, questioned the impartiality of Bellcore, or, for that matter, any "aligned" administrator. With this in mind, the FCC adopted an *industry model* to replace Bellcore's role as administrator. There are two entities. The North American Numbering Council is a quasi-governmental policy board composed of industry representatives and government observers. This council oversees a nongovernmental commercial entity called the North American Numbering Plan Administrator (NANPA). Since the NANPA must be independent, Bellcore and AT&T were eliminated from bidding on this job. The FCC considered taking over the function itself but decided in favor of the aforementioned *industry model*.[30]

The NANP is based on the "destination code" principle where each telephone number has a specific destination code or address assigned to it.[31] While the format of a telephone number has remained stable over the years, the actual allowable digits for specific portions of the format were altered in 1960, 1975, and 1995 to make more numbers available.[32] Population growth, compounded by the advent of the FAX, personal computers with modems, cellular telephones, and other new services, created demand for more and more numbers during the 1990s and into the twenty-first century. With the advent of broadband Internet access and "computerized faxing," FCC and PSC policies aimed at number conservation coupled with many individuals giving up their landline in favor of paying only for a cell phone, a few less traditional numbers are now being gobbled up. The standard telephone number format is a ten-digit number with two basic parts:

1. A three-digit Numbering Plan Area (NPA), also called an area code.
2. A seven-digit directory number consisting of:
   a. A three-digit central office (CO) code. This refers to the central office that provides the subscriber with a dial tone.
   b. A four-digit station number that uniquely identifies each subscriber within each central office code.

The original allowable digits for this format were as follows:

Area Code	Directory Number
N 0/1 X	N N X - X X X X
Where	N = any digit 2 through 9
	X = any digit 0 through 9
	0/1 = either 0 or 1
	Codes in the N11 format are excluded.

The designation of a 0/1 in the middle digit of the NPA, but not in the middle digit of a central office code, allowed for both operators and switching equipment to distinguish between a toll call and a local call and to expect ten digits rather than seven when the call was a long distance toll call. Today's allowable digits within the same format have changed this differentiation. The current format and allowable digits are:

Area Code	Directory Number
N X X	N X X - X X X X
Where	N = any digit 2 through 9
	X = any digit 0 through 9
	Codes in the N11 format are excluded.

This expanded designation added 640 new area codes to the original 152, yielding a total of 792 area code designations of the N X X format. There are a similar number of central office codes available. N11 codes are for special purposes or reserved. For instance, local directory assistance is 411 and emergency service is 911. It should be noted that the United States is very gradually evolving to a mandatory ten-digit dialing scheme for all calling. This becomes necessary in places that have adopted a total-area overlay NPA on top of the old area code rather than choosing a geographic split when number shortages require the addition of a new NPA. It's also interesting that several years ago, the FCC started the official rule-making process by initiating a Notice of Proposed Rule Making (NPRM) relative to this movement but to date never issued a final Report and Order. Perhaps too many people would complain about officially being told that they must eventually move to ten-digit dialing. This will still occur, but for now the FCC is just letting it all gradually evolve.

As one can easily see, the standard circuit-switched telephone number format and allowable digit designations are straightforward and uncomplicated. This simple number, then, is used for at least four basic functions which, it turns out, may not always be so straightforward and uncomplicated. The four functions are:[33]

- They facilitate local and long-distance telephone companies in the identification of individual customers. To facilitate and promote usage, directories are provided and directory assistance is made available.

- Numbers accumulate billing information for local and long-distance calls, operator service, and more.

- Charging for long-distance toll calls in the past was based on a time-of-day-and-distance algorithm. The calling number to called number associated rate centers were utilized to find the vertical and horizontal coordinates and to calculate the distance. This distance factor coupled with the duration (length in time) of the call was then used to calculate the actual cost of the long distance call. The industry is moving away from that approach today. Postalized charging has begun to creep into long distance charging schemes. Postalized calling is nondistance sensitive. Also, today's many and varied VoIP and cellular calling plans often offer "all you can eat" pricing packages. The bottom line is that nondistance and nontime-sensitive rates are now the norm while bundled packages are becoming the norm.

- Telephone numbers are used to route calls through the network to their destination. As mentioned earlier, this has been based on a geographic destination principle where each telephone has a specific destination code or address assigned to it.

The world is much more complicated today than when telephone numbers were first assigned to individuals to make calling more efficient. We are now a mobile society with new services and multiple numbers. We are also a world society that is experiencing the IP revolution due, initially, to the very low cost of Internet telephony, a.k.a. Voice Over IP (VoIP) calling.

Let us now turn to VoIP calling. Internet telephones use Internet addressing, which currently consists of a 32-bit (binary) numeric address in the format of four 8-bit words separated by periods. Each number would typically be represented not in its binary state, but in so-called "dotted decimal" format. If you were to direct your favorite Web browser to http://www.fcc.gov, you would be directed to their IP address, which is 192.104.54.5. If one were to convert this "dotted decimal" to binary, it would be in what we'll call "dotted binary" format "11000000 . 1101000 . 110110 . 101." Please check this for me. Point your web browser to: http://www.fcc.gov , then open another browser or another tab for your primary Web browser, and now point your browser to 192.104.54.5. This proves that the URL you are familiar with is really represented by a dotted decimal equivalent IP address.

To make a VoIP call, if you had the proper VoIP software on your PC, you could call IP address to IP address, and, using your PC headphone, you could talk–likely for free. Of course, the friends that you call would likely be represented by name, and when you call, their IP address would be sent/dialed. There are many different VoIP software and service providers: SKYPE, Vonage, Packet8, Time Warner, and, of course, more traditional telephone companies like AT&T, Verizon, and Frontier, to name just a few.

Earlier it was mentioned that IP addressing currently consists of a 32-bit (binary) numeric address, which is the basic limit for IP Version 4. This equates to 2^{32}, or about 4.3 billion addresses, which are not enough for the future when not only every person but nearly every device will be "intelligent" and need an address. In order to provide sufficient IP addresses for the future, the world will eventually move to IP Version 6, a.k.a. IPV6, which will provide for 2^{128} or about 3.4×10^{38} addresses, which we hope will be quite enough for not only the world's growing population but for every device imaginable to be directly addressable.

Now that we know how to address in both POTS and VoIP, let us direct our thoughts to calling between the old and new approach. In order to facilitate traditional (POTS) telephone calling with VoIP calling, something called ENUM was created. ENUM stands for TElephone NUmber Mapping and is a suite of protocols to unify the so-called traditional E.164-based telephone numbering system with the Internet addressing system DNS via an indirect look-up method with records stored in the DNS database.[34] ENUM was developed by the Internet Engineering Task Force (IETF), an open, all-volunteer Internet standards organization. An earlier and perhaps colloquial view of "ENUM is that it simply stands for Electronic NUMbering (ENUM) and is a protocol developed in the IEFT as RF 2916 for fetching Universal Resource Identifiers (URLs) given an E.164 number."[35]

While the movement to low–cost VoIP calling — helped by modern IP networks and addressing—is a major change to the consumer, there are other subtle changes that one should recognize:

1. Today we tend to believe that the NPA of a telephone number will help us to discern the geographical location of an address. VoIP is one of several reasons that this may no longer be true. What might be another reason?

2. Just as today we don't really care about the length of an e-mail address, embedded directories in telephone instruments similar to that found in cell phones of today will tend to make users less likely to care about the length of telephone numbers in the future.

3. In 2009, most cellular calling is still circuit-switched based. Cellular will also evolve to a VoIP system and gain efficiencies.

4. Cellular has, for some time, caused the number of landline telephones in service to diminish as many people opt for their cell phone as their only telephone. We haven't covered the reason yet, but a later chapter will indicate that while cell phones will con-

tinue to increase and wireless communications of all types will greatly advance, cellular's displacement of landlines will diminish in the future. Stay tuned for the reason for this phenomenon.

5. Existing circuit-switched networks like the PSTN will likely remain in service for some time due to the fact that they work well and much of the sunk capital is already depreciated.

The study of numbers and addressing is a truly interesting part of U.S. and global telecommunications networks. While the powerful Google has revolutionized the find/search activities of humankind, and this technique may eventually eliminate the need for today's traditional directories, many still feel that only through proper addressing will person-to-person communications remain as easy and efficient as it has been thus far.

Time will tell.

Toll Free Service

In the 1960s, AT&T recognized a business marketing need. Large businesses were receiving collect calls from their customers and suppliers via the operator-handled, collect-call process in order to shift the substantial long-distance call cost away from customers, thereby allowing customers to have easier, low-cost access to them. This collect-call process was time consuming for Bell and expensive for the business. Bell sought an improved solution. Before we learn about this *improved solution*, it's important to stop and discuss the various types of calling that were prevalent in the past, because if you are reading this and were born after 1970, it's very likely that you may not understand the difference between a collect long-distance call as compared to what we think of today as a traditional long-distance call. Telephone people tend to refer to traditional long-distance calling as a DDD or a Direct Distance Dialing call or merely a sent-paid call. The term sent-paid is an old term that was used because it is analogous to when you mail a letter and put a stamp on it–you the sender pays for it. A collect call is paid for by the receiving party. If you were born after 1970, it's likely that you didn't make your first long-distance telephone call until sometime in the early-to-mid 1980s and, by then, very few people in the United

Telephone Operator, circa 1900[36]
Photo Courtesy of the Richardson, Texas
Historical and Genealogical Society

U.S. Air Force Sgt. operates a switchboard
in the underground command post at
Strategic Air Command headquarters,
Offutt Air Force Base, Neb., 1967[37]
Photo compliments of the U.S. Air Force

States made or experienced collect calls. In fact, few people living in
the United States today that are younger than forty years old ever experienced operator-assisted type of telephone calls. So, the younger you are,
the more likely you have no idea of what a collect call was/is, so it's well
worth taking just a moment to address types of calling. In the early days
of the telephone, most calls were placed from an individual to another
by way of an operator who actually physically made the connection via
a cord board. Above are two photos of old telephone "cord boards."

Operator-assisted calling was very time consuming and labor intensive and was eventually replaced for most so-called sent-paid calling
by Direct Distance Dialing, also known as DDD calling. The first Direct Distance Dialing occurred in late 1951 in New Jersey and quickly
spread throughout the land. Direct Distance Dialing required an automatic switch that could connect the calling party to the called party
without operator intervention. Prior to DDD, calls were all operator assisted. Direct Distance Dialing allowed callers to quickly and efficiently
place long-distance calls themselves, without any operator assistance.
DDD was/is much less expensive than operator-assisted calling. Yet, operator-assisted calling is still available to you today from your landline
circuit-switched telephone, although it is a bit costly. If one were to visit
the beginning pages of their local telephone directory, one would learn
that in addition to the cost of the call, there is an additional fee for any
collect, time and charges, and operator-dialed calling card calls. Roch-

ester, New York, is served by Frontier Telephone and their additional charge is $1.25. Third-number calls, which allow one to call from their landline to another number and have it charged to a third-number, are even more costly. Please check this out by investigating the front of your own telephone directory.

In 1967, AT&T launched toll-free 800 services as the improved solution to address the aforementioned needs.[38] The service was initially viewed as a technological convenience with limited appeal, but due to excellent public and business acceptance, coupled with advances in both technology and business marketing techniques, toll-free service grew into an over $8 billion per-year revenue stream by the mid 1990s and over $10 billion into the twenty-first century. Today developments in online services directly compete with 800-type call centers, and mobile and VoIP packaged call plans have created a situation that has slowed the tremendous growth of toll-free services. Let us look at the path that toll-free service has taken.

From the user's viewpoint, if a large user deployed 800 service for their clients to contact them, the client need merely call a ten-digit number beginning with 800 and the call could connect without charge to the called party. Early in the product life cycle of toll-free service, a business that ordered the 800 service was charged under the Inward Wide Area Telecommunications Services (INWATS) tariff, which consisted of a large monthly fee for the 800 number and a reduced rate per call. For example, a company might pay $5,000 per month for the 800 service number and $.04 per minute, rather than $.20 per minute. Today most toll-free services for large clients are bundled with outbound services and provided as a package under special tariffs. Sometimes data services are also included in these bundled offerings, and no large per-month charge is incurred and the cost per minute after all discounts would normally be less than $0.10 per minute—often much less.

AT&T invented the service and technology for toll-free calling and the network architecture to support it. From 1967 until about 1981, routing and screening for the 800 calls was basic, and there were many regulations and restrictions. Since there were different regulations and tariffs for interstate versus intrastate calling, different numbers had to be used for in-state calls versus interstate calls. Another limitation occurred if a subscriber of the toll-free service moved to a new location or wanted to increase the area covered by its 800 number. The subscriber usually would have to change 800 numbers or even have multiple 800 numbers and also could not choose the 800 number that it wanted.

In 1981 a major technological advancement came to the network in the form of Common Channel Interoffice Signaling (CCIS). CCIS is a separate and distinct, special and very fast signaling channel between telephone central offices that is different from the voice channel or path and is used to set up the voice or talking path. CCIS offers basically two benefits. First, it dramatically speeds up the setting up and tearing down of a phone call. Second, it allows for much more information to be carried about the call than was carried on in-band (old-fashioned) signaling.[39] This advancement, coupled with the introduction of a first-generation system of centralized databases for processing 800-number calls, provided several new toll-free service features and benefits.[40] Customers could choose their own 800 number, often called a "vanity" number. For instance, a florist might choose 1-800-FLOW-ERS and thereby provide an easily recognizable and remembered number. More business would hopefully be the end result. Some additional new features were that the geographic coverage area became variable, clients could move without changing their number and calls could be routed to different locations based on factors such as time of day, day of the week, call place origination, and overall calling volume.

The next major relevant event was announced in 1982, and it was the Modification of Final Judgment[41] (MFJ), which we more commonly call divestiture or the breakup of the Bell System. Divestiture took place on January 1, 1984. Among other things, this action placed the local telephone service and intraLATA 800 service with the Regional Bell Operating Companies (RBOCs), and interstate long-distance service, including interstate 800 traffic, went to AT&T. Other interexchange carriers (IXCs) were competing for long-distance service, but AT&T remained the sole interstate 800-service provider until the latter part of the 1980s. In 1985 Judge Greene mandated that RBOCs offer exchange access for 800 service to all IXCs, not just AT&T as had been the practice. That decision led to an "interim 800-NXX plan," which was administered by Bellcore. Bellcore allocated unique 800-

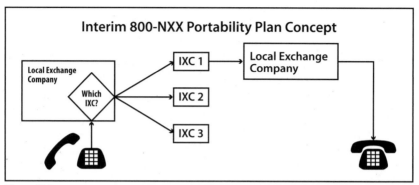

Figure 2.3

NXX codes to each IXC and Local Exchange Company (LEC) that offered interLATA and intraLATA 800 services. Local exchange carriers screened each originating 800-service call at the exchange level and routed it to the appropriate carrier based on table look up relative to NXX. Once routed to the proper IXC, the IXC did its own POTS number translation and service coverage verification.[42] This interim plan basically allowed each carrier to provide 800 service, but each carrier could only offer the service based on the Bellcore allocated NXX codes. For a customer to move from one carrier to another, it would have to change 800 numbers. Figure 2.3 indicates how an interstate 800 call would route and flow to completion under the NXX screening plan.

After three years of investigation and debate (CC Docket 86-10), in 1989 the FCC concluded that it was in the public interest for LECs to provide *equal* access for 800 service. The FCC directed the LECs to deploy *equal* access after they deployed enough Signaling System 7 (SS7) capability so as to not degrade service access times. Access time was also more specifically referred to as post-dial delay and was a major industry concern at the time. The FCC required that most of the 800 service traffic be protected from additional post-dial delay prior to moving to the final 800 number portability plan. Under the three-digit NNX interim plan, the time required to process the three-digit NNXs ranged from 1.9 to 3.5 seconds. The FCC mandated that 97% of the access traffic should be processed and reach the correct IXC within five seconds and that by March of 1995 the mean access time should be 2.5 seconds.[43]

While AT&T's competitors were anxious to begin competing for the 800 marketplace without the obvious drawbacks of the interim 800-NXX plan, updating all central offices and implementing the FCC quality-of-service mandates were very large problems for the Local Exchange Companies. As a result, the industry challenged the FCC decision and argued that those parameters might hold up the 800 database access indefinitely and that a more market-driven quality-of-service standard should be used to move to the new 800 database. This situation persuaded the FCC to relax its original criteria and to expedite the deployment timetable. Testing showed that actual access times could not always be achieved in less than five seconds. This was true when multifrequency in-band signaling was involved Please see Table 2.1.

Table 2.1 Actual vs. Theoretical Access Times[44]

Configuration	Theoretical Time	Low Tested Time	High Tested Time
EO>SSP>SS7>IXC	0.8	0.5	0.9
EO>SS7>AT/SSO>SS7>IXC	1.0	0.6	1.4
EO>MF>AT/SSP>SS7>IXC	4.1	4.2	6.3
EO/SSP>MF>IXC	4.8	4.4	5.6
EO>MF>AT/SSP>MF>IXC	6.4	7.1	9.4
AT Access Tandem **EO** End Office **IXC** Interexchange Carrier POP **MF** Multifrequency In-band Signaling **SSP** Service Switch Point **SS7** Signaling System 7			

Due to actual testing and an analysis of the specific technological makeup of their networks, the various RBOCs and major independent telephone companies asked for waivers. Table 2.2 shows their individual requests.

Table 2.2[45]

800 Database: Status of RBOC/Telco Waivers to Access Time		
RBOC/Telco	**March 1993** **(97%< 5 seconds)**	**March 1995** **(100% < 5 seconds)** **(@ 2.5 seconds mean)**
Ameritech	Waiver Requested (78%)	Waiver Requested (94%)
Bell Atlantic	Waiver Requested (85%)	Will Comply (100%)
Bell South	Waiver Requested (93%)	Waiver Requested (98%)
GTE	Waiver Requested (90%< 5 seconds, 97% < 5.5 seconds)	Will Comply (100%)
NYNEX	Waiver Requested (92% < 5 seconds, 100% < 6.5, and 97% < 5.5 seconds)	Waiver Requested (98% < 5 seconds, 100% < 6.5 seconds)
Pacific Bell	Waiver Requested (97% < 6 seconds)	Waiver Requested (2.3% > 5 seconds)
SW Bell	Waiver Requested (72% < 5 seconds	Waiver Requested (97%)
U.S. West	Waiver Requested (64% 5 seconds, 82% < 6 seconds, and 91% < 6.8)	Will Comply (100%)
United Tel	Waiver Requested (90% < 5 seconds, 97% by October)	Waiver Requested (Delay to Dec 1997)

Percentage numbers in each column reflect the percentage of 800 traffic estimated to meet or exceed the corresponding FCC requirements.

The FCC denied all waiver requests relative to the 1995 standards but did waive some of the March 1993 standards with the condition that SS7 be expedited. LECs were given until March 1993 to fulfill the FCC's mandate. Toll-free calling and major sales and marketing call centers are real cash machines. Each successful sale means revenue to the business. The toll-free 800 market had grown to about $7 billion by 1992. As a result of the tremendous money at stake, coupled with worry, in part due to the aggressive LEC implementation schedule, and also to the potential increase in lost or abandoned calls that might occur as a result of a potential increase in post-dial delay, an ad hoc committee of 800 users successfully lobbied the FCC in early 1993 to postpone the LEC deadline from March to May 1993 in order to provide the industry more time to get insure success.

The introduction of the 800 database solution to 800 portability proved to be technically excellent despite pre-portability concerns and anxiety. Brian Brewer, director of 800 marketing for MCI at the time, said the transition to portability had been "extremely smooth," and further indicated that "Call setup time is actually better, due to the implementation of SS7."[46] Cedric Smith, national marketing director of 800 services for AT&T said that "the industry has shown a great deal of cooperation."[47] Much planning, investment, coordination, and installation occurred in preparation for the May 1, 1993 database portability for 800 numbers. From the 800 subscribers' perspective and that of their callers, it was a major non-event and that is just the way they wanted it.

Just how does an 800 call occur under the 800 number portability database plan? When an individual dials the "800" in an 800 number, it alerts the service switching point (SSP) associated with the end or tandem central office from which the individual call originated. The SSP is a system that provides part of the intelligence in intelligent networks. It recognizes calls that require special handling and can communicate with other intelligent-network systems, such as service control points (SCPs), to obtain and implement call-processing instructions. The following is a detailed narrative of 800 call processing:

> Upon recognizing that a call is to an 800 number, the SSP contacts the appropriate SCP to obtain detailed instructions for processing the call. The SCP finds this information by checking the customer record identified by the 800 number in the data message received from the SSP. The necessary instructions are then sent back to the service switching point to be carried out. This could entail completing the call within a LATA (local access and transport area) according to an 800 service customer's instructions in the SCP database used by the local exchange company or the call might be forwarded to an interexchange carrier, with subsequent routing determined by the customer's instructions in that carrier's 800 service database.[48] (See figure 2.4)

The Service Management System, or SMS/800, displayed in figure 2.4, is a software system originally developed by Bellcore and is used to create and maintain the SCP records. Every 800 service provider that assigns 800 numbers must use the system to verify the availability of specific numbers and reserve them, as well as create customer re-

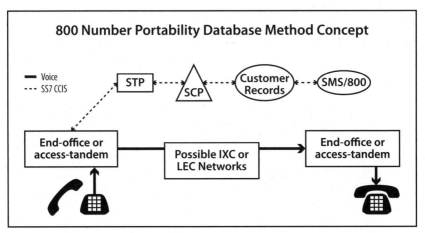

Figure 2.4

cords and download records to local exchange carrier SCPs to facilitate call processing. Bellcore formed Database Service Management, Inc. as a wholly-owned subsidiary in April 1993 to operate and maintain the SMS/800 system.[49] The term "Responsible Organization," or "RESPORG" for short, was adopted by the telecommunications industry to identify a customer's primary point of contact for a given 800 number. Only this designated RESPORG party is authorized to interface with DSM and make changes to the SMS/800 database.

It is difficult to assess the exact market effect of 800 portability on major carriers since all claimed success. AT&T had the lion's share of the 800 market with over an 80% share, compared to MCI at less than 10%, and, therefore, AT&T had the most to lose. Yet in late October 1993, AT&T reported that service orders for 800 service more than doubled since portability. AT&T signed and added more than 140,000 new 800 numbers to its customer base, representing better than $500 million in annualized revenue with 63,000 of those new numbers coming directly from its competitors, and minutes of use on AT&T's 800 Network were experiencing double-digit growth.[50] At the same time, MCI and Sprint reported market share gains of 4% and 2% respectively. Possibly this discrepancy may reflect increased 800 traffic stimulated by the advent of portability, as well as the introduction of new vertical service capabilities for 800.[51] The relatively new marketing of 800 service to residence customers also could have been a factor. The discrepancy in market share claims may be due to the dynamic expansion of the market plus marketing hype of some or all of the major carriers. One should remember that few public relations efforts improve public opinion as much as success-oriented ones. Often the perception of success helps to yield success. Reported increases in market share could only help public relations.

Partially as a result of the expedited introduction of SS7 to the network, many new services and features were introduced. These include the ability to split 800 traffic by carrier, the ability to provide customer-controlled programming, interactive voice response, caller-recognition routing, and advanced ACD-like features, such as next-available agent. Also, quality-type features and functions like AT&T's "Never Miss a Call Guarantee" or "Five Minute Service Assurance," and Sprint's "Sprint Secure" and MCI's "MCI 800 Guardian" were introduced. Few, if any, carriers will publish how much their effective rates (the end cost after all discounts are applied) to clients have gone down. However, it is generally felt that due to the increased competition caused by 800-number portability, rates have diminished. According to Robert A. Gamble, senior telecommunications analyst responsible for multiple inbound call centers at the time, "IXC's marketing philosophies revolve around four issues. They are: network reliability, competitive packages and rates, service, and network features."[52] It would appear that as a result of the expedited introduction of advanced technologies like SS7, as well as the increased competition of major 800 providers, each of these marketing issues was impacted by 800-number portability. Moreover, these impacts were positive to both purchasers and users of toll-free 800 services.

Another interesting event took place in the mid-1990s commonly referred to as 800 number exhaust. In the spring of 1995, the supply of unused 800 numbers dropped below 600,000, so the FCC stepped in and ordered 800-number rationing until the Public Switched Telephone Network (PSTN) could accommodate the alternative toll-free number blocks. In January 1995, it was determined that new toll-free number blocks consisting of the prefixes 888, 877, 866, etc., could be used as toll-free calling just like 800. However, LECs were given time to update their central offices to recognize the 888 toll-free prefix.

When 800-number portability went into effect in May 1993, three of the eight million available numbers were in service. Fourteen months later, more than 7.4 million 800 numbers were in circulation, and the consumption rate was running at about 113,000 per week.[53] This was partly due to large blocks of numbers being warehoused. Hoarding and brokering of 800 vanity numbers by middlemen had occurred. The FCC recognized these problems and initiated steps to improve the situation via CC Docket No. 95-155, which specifically prevents warehousing and intends to promote efficient use and fair and equitable distribution of toll-free numbers. The toll-free 888 prefix went into effect in the spring of 1996 and has provided relief to the 800 exhaust situation. This is one more indication of the growth in toll-free calling and the value that some people placed on vanity numbers in the 1990s.

AT&T created toll-free 800 service and introduced it in 1967 during a time when the economic model of the telecommunications industry in the United States was slowly evolving from a monopolistic model toward a competitive model and the Public Switched Telephone Network (PSTN) was deploying the circuit switch technology of the day. The divestiture of the Bell System had no direct effect on toll-free calling, but afterwards, in 1985, Judge Greene mandated that the RBOCs offer exchange service to all IXCs, not just AT&T so as to introduce competition into toll-free calling. This decision led to an "interim 800-NXX plan" for 800 numbers. In 1989 the FCC decided that it was in the public interest for LECs to provide *equal* access for 800 service—not just to AT&T. This decision facilitated the introduction of SS7 into the network and total toll-free 800-number portability in May 1993. Note that first came the decision to allow 800 access service in 1985, then the interim 800-NXX plan, followed by the 1989 FCC decision requiring equal access for 800 service and, finally, true 800-number portability in May 1993. Equal access and toll-free competition took a while.

From 1967 until today encompasses over forty years of advances in technology, policy and legal movement to a much more competitive economic model. The quality and reliability of the network significantly improved as demonstrated by carrier service guarantees for 800 service. So, too, did the breadth and depth of 800 services available to clients. Volumes of toll-free calling climbed through the 1990s while per-call prices for toll-free service for corporations fell, in part due to competition. AT&T's competitors likely felt that it took far too long for them to be able to compete in this market.

More than twenty-five years after divestiture, toll-free service is no longer enjoying tremendous growth. What was once an over $11 billion U.S. revenue stream will likely decline. Toll-free service was created based upon the premise that a long-distance call was costly and businesses wished to facilitate customer access without the burden of cost to their customers. Today, competition and VoIP have reduced the cost of long-distance calling. Further, bundled VoIP and mobile plans provide the perception of "all in one" calling plans and no-cost long distance. Toll-free service today is somewhat like an individual who is outside in a rainstorm with an umbrella and raincoat. Once the sun comes out, the value of the rain gear diminishes although it sometimes takes time for that individual to realize that it stopped raining. The value of toll-free services seems headed the way of the raincoat and umbrella after the rain stops. Today many are slowly realizing that the rain has stopped. The day may come when there is no value and no revenue stream for toll-free services.

The following is an article that succinctly reinforces this concept. Its source is the USA Today article by Kevin Maney.[54] It is reprinted here with permission from USA Today.

A glance into the crystal ball hints at a future without 800-numbers

The 800-number — for 40 years a part of daily American life — is doomed.

Like what happened to pay phones. And milkmen.

This would be very bad news for phone companies, which rake in $12 billion a year from toll-free numbers.

The 800-number's destiny first occurred to me a couple of months ago as I stood outside a neighborhood hardware store looking at Weber gas grills. On each grill was a sticker that said if you have any questions, call this 800-number.

Clever, right? Just about anyone who is out looking at Weber grills is probably carrying a cell phone. And a Weber call center person no doubt can explain the grill better than a part-time hardware store clerk, assuming the Weber call center person is not a subcontractor in Mumbai who is also taking calls on the other line about doggy diapers.

Except there's something odd about this equation. Just about everyone who has a cell phone has a flat rate package for local and long-distance calls. In other words, as I stood there with my phone, there really would've been no difference whether I called a toll-free 800-number or a "toll" 847 area code number at Weber's headquarters in Palatine, Ill. Both calls would've cost me essentially nothing.

But if I call Weber's 800-number, the call costs Weber at least a few cents a minute. Those calls add up to millions of dollars a year for a company like Weber.

Huh, I thought. Why would a company spend all that money it didn't have to spend?

That lingered until, a couple weeks later, Mike McCue, the CEO of Tellme, came by our offices. Tellme makes many of the voice-recognition systems you might run into on the phone — such as when you call 411, or an airline, or Domino's Pizza, and get a computer-generated voice that tries to help.

Tellme's next generation of services, McCue says, will be through a single button on mobile phones. So instead of dialing 1-800-DOMINOS to find the nearest pizza outlet, you'd push a button on the phone and say, "Pizza." And then the phone's screen would show a list of pizza places in your area. Click on one and the phone would dial it. If you just said, "Domino's," the phone would show the nearest Domino's and dial it.

That kind of service will be another reason a business, such as Domino's or Weber, would not need to spend money on an 800 number. If I'd had one of these phones — which are going to start appearing at the end of 2006 — at the hardware store, I would've just pushed the button and said, "Weber." I would've been connected to that nice grill expert in Mumbai.

Tellme's type of service — and certainly others will offer similar services — would also threaten the thriving industry of vanity 800 numbers, such as 1-800-PUP-POOP, which belongs to a company called Scoop Masters, "Tampa Bay's premier dog waste removal service." (Sorry, I seem to be in a pet waste sort of mindset.)

"If you can pick up the phone and say, 'Flowers,' it breaks the 1-800-FLOWERS issue," McCue says.

He ticks off all the forces lining up against 800-numbers. It's not just cell phones. Most home phone plans these days bundle local and long-distance calling into a flat rate, so calling an 800 number from a home phone isn't any more free than calling a regular number. Same thing with the growing number of Internet calling plans.

Then there's the Web. Calling an 800-number and getting routed into voice menus or waiting forever for a live person has persuaded a lot of people to go to a company's Web site first. And companies are finding better and cheaper ways to connect with consumers right through their Web sites, like with IM-style chat help or "push to talk" buttons from Skype and others.

At some point, this will reach a tipping point. Companies will decide they no longer need an 800-number to allow the vast majority of consumers to reach them. When you add it all up, the toll-free-number industry "is just going to collapse," McCue says.

Now, it's important to note that this hasn't happened yet, and not everyone agrees with McCue. A just-out report from Insight Research says that "Web-enabled customer services are not displacing toll-free customer services." A graph shows toll-free revenue increasing slightly through 2009.

The analysts at TNS Telecom have a little less sanguine view. They note that the average amount spent monthly by businesses on toll-free numbers dipped about 2% from 2004 to 2005. However, TNS Senior Vice President Charles White says, "Despite the varied ways for firms to receive inbound communication, toll-free service continues to be a trusted option for many firms."

Which, I'm sure, is true. There was also a time when an analyst could've said, "Despite the introduction of gasoline-powered tractors, horse-drawn plows remain a trusted option for many farmers."

No doubt the 800-number — introduced in 1967 — will have its place for a while. The forces coming to bear on it are still in a reasonably early stage.

Plus, the 800-number has one distinct advantage: its hold on the consumer psyche. "It does indicate 'free,'" McCue notes. If the number to call on the Weber grill had started with 847 instead of 800, would I have been as willing to dial it? Maybe not. Or at least I would've thought twice. And a company like Weber can't yet take the chance that I might not call because I think the connection will cost me.

But give it time. It will sink in for the majority of us that all calls are the same cost. And businesses will start asking why they're paying these toll-free-line bills or renting a vanity number from some firm such as Dial 800 — a company that licenses 1-800-DUMP-STER to Waste Management for $99 a month and eleven cents per minute, per call.

Then a $12 billion industry is going to slip the way of radio dramas.

From USA Today, a division of Gannett Co., Inc. Reprinted with permission.

U.S. Telecommunications Policy

Let us now turn our attention to telecommunications policy, which includes telecommunications regulation and law. From chapter one, you already know the meaning of telecommunications. The term "policy" generally means a set of principles, perhaps a plan of action, to be used in making decisions. Typically, public policy, which encompasses telecommunications policy, is set in order to attain some goal or outcome. Often, as in the case of much telecommunications policy, the goals are based upon social values. So, then, telecommunications policy is a set of rules or guidelines directed to goals and outcomes that society, as represented by the policymakers, deems beneficial based upon the values and goals of society. Societal goals may change over time and then one

Policy is supposed to help meet societal goals

can sometimes observe changes in policy and desired outcomes. An example is the shift from an accepted and protected telecommunications monopoly model for most of the twentieth century to the more competitive model of this twenty-first century.

Now that we understand the term *telecommunications policy*, it's worth addressing the question: Why understand and study telecommunications policy? Early in chapter one, it's mentioned that our telecommunications environment was supported/affected by four main factors. They are: technology, market forces, public policy, and security. The invention of the telephone, loading coil, transistor, fiber optics, and IP networking are all examples of technological advances that have greatly affected the telecommunications environment. For now, let's accept market forces and security on faith and keep our focus on telecommunications policy. The term *information society* came into everyday use some years ago when economists and others used it to describe the period of time when a society moved from one based on material goods to one based on knowledge. Now more than ever we are an information society whose information is provided to mankind via telephone, Internet, TV, and other forms of telecommunications. Sadly, books and even the printed newspaper have much less of an impact on society today than during the twentieth century. The very positive flip side is that telecommunications plays an even greater role. So, telecommunications influences who we are intellectually as much as what we eat influences what we are physically. Please remember that genetics, the environment, and lifestyle factors also influence what we are physically. Your intellect is greatly influenced by the many communications you receive, and most are via telecommunications. Also, telecommunications policy impacts the telecommunications environment so as to impact us and the industry as users, some corporations as providers and telecommunications workers and professionals in their careers. While not as self-evident as the technological advances outlined above, the following are three telecommunications-policy initiatives that had great impact on the telecommunications environment:

- The Kingsbury Commitment stopped the Bell System from "gobbling" up their competitors and forced interconnection and cooperation.

- Divestiture of the Bell System broke up the Bell System and created a marketplace that enabled long-distance choice, competition, and greatly lowered user long-distance costs.
- The Telecom Act of 1996 "enabled" competition in all other areas of telecommunications including local service. It allowed all "players" to compete in all markets.

The importance of telecommunications policy leads us to even more questions. Before we address them, let's recap some things that you should have already learned regarding U.S. Telecommunications Policy:

- Unlike most of the rest of the Western world, the U.S. telecommunications industry grew up in the private sector. What significant event from Chapter 1 was the pivotal issue to make this occur?
- U.S. patent rights led to a period of no competition during the early years of the telephone. This was followed by a period of intense competition, which again was followed by a period of much less competition when the patented loading coil enabled AT&T to offer long-distance service. The benefits bestowed upon a patent holder are a form of U.S. Public Policy.
- Theodore Vail's concept of Universal Service was a social policy originally directed to provide a subsidy to rural individuals for basic telephone service by slightly overcharging municipal-based customers. This helped individuals to have universal access to telecommunications and it helped AT&T's network to grow.
- The Kingsbury Commitment was an out-of-court settlement between AT&T and the Department of Justice (DOJ). In addition to halting AT&T's purchase of independent telephone companies and forcing AT&T to interconnect with other telephone companies, it established a partnership between Bell and its competitors, and it fostered cooperation between AT&T and the government. It also was the first of many "out-of-court" settlements.
- The first major telecommunications legislation was the Telecom Act of 1934. This created the FCC.
- In 1949 the DOJ filed an antitrust suit attempting to breakup

the Bell System. This was settled by the 1956 Consent Decree, which did not break it up but provided a nonexclusive license and access to all technical breakthroughs of the Bell System, which eliminated any possible technological supremacy of the Bell System.

- On January 8, 1982 the DOJ and AT&T agreed to breakup the Bell System. This was divestiture a.k.a. the Modified Final Judgment (MFJ). The modification refers to the changes made to the 1956 Consent Decree.

The major questions that one should address before continuing are:

- Who makes telecommunications policy?
- What are the processes by which they do so?
- Who do we regulate?
- Why do we regulate?
- How do we regulate?

Who makes telecommunications policy? In order to answer this, let's first take that dangerous step and discuss law from a layman or nonattorney's viewpoint. Please note that these are personal views and not necessarily those of the U.S. Government or any legal entity, attorney, or law school. The U.S. law system is based upon four sources of law. They are: Constitutional Law, statutes, administrative regulations and decisions, and common law. First and most important is Constitutional law. The U.S. Constitution decrees the organization of the U.S. government as well as duties and specific individual rights of citizens. In addition to the federal Constitution, states also have state constitutions. Together, this yields a system of government whereby the federal government has certain responsibilities and states have separate and distinct responsibilities given to them. It is important to note that states' rights indicate that for those responsibilities entrusted to them, they are NOT subordinate to the federal government. Also, within the framework of the U.S. Constitution, we have a system of three separate branches of government with separate and distinct responsibilities and powers. The three branches are Executive, Legislative, and Judicial. The Executive Branch includes the President and is responsible to insure that laws are carried out. The Legislative Branch includes both the House of Representatives and the U.S. Senate and is responsible for making laws. While they debate and craft a bill, the bill does not become law until it is passed by the President. The Judicial Branch interprets laws and makes sure that they are in line with the Constitution — the primary law of the land. The Judicial Branch also insures that laws are applied in a uniform and consistent manner and applied without discrimination of any kind. Together, this separation of power yields what our founding fathers considered to be a system of checks and balances. This helps to guarantee that no portion of the government becomes too powerful.

The second source of law is statutes. These are laws passed by Congress or state legislatures or city governments. An example would be the Telecom Act of 1934. Third would be administrative regulations and decisions. An example would be FCC telecommunications decisions. These, like all laws, are "appealable" via the judicial system. Fourth and last would be Common Law, which is used where no statute or law addresses an issue. Ideally, precedent is used to attempt to yield a more consistently applied remedy.

Knowing the four sources of law helps us to understand some policy issues. For instance, after years of not addressing the vast changes in the telecommunications arena, Congress passed the Telecommunications Bill of 1996, and after it was signed into law by President Clinton on February 8, 1996 it became the Telecommunications Act of 1996. Also, we have the FCC that was created by the Telecom Act of 1934 and directed toward interstate federal telecommunications issues with each state having a Public Utility Commission (PUC) or Public Service Commission (PSC) directed to in-state matters. Therein lies the reason for having both interstate telecommunications tariffs and state-wide telecommunications tariffs. This is not unique to the telecommunications industry. There are federal and state rates for the moving industry as well. If you moved your home furnishings from Rochester, New York to New York City, you would be charged under statewide rates, but if you moved from Rochester, New York, to Boulder, Colorado, you would be charged under interstate rates. So, who makes telecommunications policy in the United States? Basically, many individuals and entities make telecommunications policy. Every citizen who votes for senators

[handwritten: wide combination of people decide telecom policy, including voters indirectly]

and representatives and a president indirectly helps to make policy. Congress crafts bills and the President signs a bill into law, while the courts interpret these laws. The Telecom Act of 1996 was more of a visionary framework or outline of goals and specifically directed the FCC to craft the details. State PUCs oversee, regulate, and set telecommunications policy for in-state issues. So, as you can see, within the United States many individuals and entities make telecommunications policy.

What are the processes by which they do so? You already have an idea of how laws are made. Here, taken from the FCC Web site, is the typical process[55] for crafting policy regarding a specific subject within the FCC's jurisdiction:

- **Notice of Inquiry (NOI):** The Commission releases an NOI for the purpose of gathering information about a broad subject or as a means of generating ideas on a specific issue. NOIs are initiated either by the Commission or an outside request.
- **Notice of Proposed Rulemaking (NPRM):** After reviewing comments from the public, the FCC may issue a Notice of Proposed Rulemaking. An NPRM contains proposed changes to the Commission's rules and seeks public comment on these proposals.
- **Further Notice of Proposed Rulemaking (FNPRM):** After reviewing your comments and the comments of others to the NPRM, the FCC may also choose to issue an FNPRM regarding specific issues raised in comments. The FNPRM provides an opportunity for you to comment further on a related or specific proposal.
- **Report and Order (R&O):** After considering comments to a Notice of Proposed Rulemaking (or Further Notice of Proposed Rulemaking), the FCC issues a Report and Order. The R&O may develop new rules, amend existing rules, or make a decision not to do so. Summaries of the R&O are published in the *Federal Register*. The *Federal Register* summary will tell you when a rule change will become effective.

Who do we regulate? The answer is that we tend to regulate firms that offer essential products or services to the general public and that have monopoly or market power over their offerings. The roots of this answer are found in both the field of economics and common law. Specifically, in the case of Munn vs. Illinois in 1877, one outcome was the two-part test: Is the service essential and does the provider have market power? If a telecommunications service meets both parts of this so-called "two-part test," then it would likely be regulated.

Why do we regulate? We regulate in order to protect individuals and firms against the exercise of market power in the provision of essential services. Similarly, the food industry is regulated in order to protect the general public from health issues. No one wishes to eat contaminated food.

How do we regulate? Typically, in the United States the vision is to limit excess profits by constraining prices/profits and discourage discrimination by various means including requiring public tariffs. This often takes place by: controlling market entry and exit, setting prices (and quality) in order to meet limits, controlling discrimination (between like customers), and limiting profit/revenues to reasonable levels. There were also times when so called "dominant carriers" were forced to sell services to their competitors at wholesale prices set by regulators in such a way as to create more competition at the retail level. Of course, more competition usually yields more consumer choices.

So far, we have discussed federal and state policy in the United States. There may also be local policy directed toward the telecommunications environment. Perhaps the best example of this would be the zoning process and laws as they relate to cell towers. Most municipalities have property zoning laws and processes in place that guide what type, size, etc., a structure may be relative to a specific location. Requirements for residentially zoned areas are more stringent than in commercial or industrial zones. Generally, the U.S. public desires the use of their cellular set, but seldom does an individual desire a cell tower in their back yard. Federal law requires municipalities to allow cell towers but also allows local municipalities to help determine the physical, visual, and other characteristics of the structures. Tower height is usually an important and sometimes controversial issue when a company desires to build a cell tower.

Concepts

There are a few subtle and even simple concepts that are important to learn in order to gain a complete understanding of the full essence of the telecommunications environment and the changes that have occurred and may still occur.

The first of those concepts is the concept of an *externality*. Webster's defines externality as "a secondary or unintended consequence."[56] Typically, the pollution example is used to illustrate this concept. A company might manufacture a product and through the manufacturing process, might pollute the environment. This of course would be a negative externality. The field of telecommunications offers many nice and positive illustrations of externalities. Let us confine our telecommunications networking thinking for the moment to only voice calling. If you were the only one on a network, it would not be of much use to you since you could only call yourself. However, as the network expands, one can call many more individuals and if there are many, many endpoints (individuals) on the network and you are capable of calling them all, then whenever an individual is added to that network, the value of the network increases and this would be a positive externality. In fact, Theodore Vail's concept of extending the early telephone network to everyone even if they were way out in the very rural farm country by slightly overcharging all the close-in city dwellers to subsidize adding a farmer is a great example of the early Bell System helping the farmer and themselves while at the same time adding value to the subscribers network.

The next interesting concept is one that might normally also be perceived as a positive externality. The term is *phatic* and Webster's Online Dictionary defines phatic as "of, relating to, or being speech used for social or emotive purposes rather than for communicating information."[57] There was a time that using the telephone was primarily for the purpose of communicating information. Today that is not the case. Just look across any college campus and see the many, many college students who are very often calling for social or emotional purposes. There is a nice song that's available for you to "*Google*" online that will explain this phenomenon well. The song is by Stevie Wonder and it's called: "I just called to say I love you."

The last major concept that we'll discuss here concerns the *eco-*

nomic regulation of telecommunications and pricing concepts and approaches. This is part of telecommunications policy. Throughout most of the history of telecommunications in the United States, telecommunications companies were required to file tariffs for their regulated products and services. Newton's Telecom Dictionary defines *tariff* as "a document filed by a regulated telephone company with a state public utility company or FCC. The tariff — a public document — details services, equipment, and pricing offered by the telephone company (a common carrier) to all potential customers. Being a 'common carrier' means it (the phone company) must offer its services to everybody at the prices and at the conditions outlined in its public tariff."[58] Typically, a telephone company would define their tariff, publicize it, and perhaps offer public hearings and then seek approval through either their state PSC or the FCC depending on whether the service was intrastate or interstate in nature. Today, some telephone company products and services are no longer regulated so they do not have a tariff. It is expected that such competitive products and services would be self-regulated by the free market economy.

For the better part of the twentieth century, a local exchange telephone company was regulated under what is called *Rate of Return regulation*. Perhaps the most succinct method of learning about this would be to offer a formula. Please note that this formula is over-simplified but it will help you understand the concept of Rate of Return regulation very well.

Assume the following terms:
- **RR** stands for Revenue Requirement
- **E** stands for Expenses
- **r** stands for Rate of Return allowed and is expressed as a percent
- **C** stands for Capital Invested
- **D** stands for Accumulated Depreciation of that capital.

Then the formula would be: **RR = E + r (C – D)**

Let us now apply this formula to the following problem. A small telephone company has only one service and that is basic dial tone. Assume that their state PSC is allowing a 7% rate of return. The company has $3 million invested capital and accumulated depreciation of $1 million. This small telco has a total anticipated expense of $980,000 for the year. If there are 10,000 subscribers to this dial tone service, what monthly charge per subscriber would be allowed?

Using the formula $RR = E + r(C - D)$ and substituting the above variables yields: $RR = \$980,000 + .07(3,000,000-1,000,000) = \$1,120,000$ **per year.** Since there are 10,000 subscribers and there are twelve months per year we would divide the $1,200,000 by 10,000 subscribers and that by twelve months per year. This two stage calculation yields an answer as to what the overall revenue requirement for this telephone company would be per year. From that, one can calculate the final answer of $9.333 per month, per customer.

The major failure of Rate of Return regulation is that there is no incentive for the telephone company to become efficient. If expenses rise, these expenses are just driven back to the rate paying public. A portion of the invested capital is also driven back to the rate-paying public. Accordingly, regulators in the latter part of the twentieth century began to look for a better approach. The better approach is something called *incentive based regulation* and attempts to incent the provider to become more efficient. There are several "flavors" of incentive based regulation. Let us conceptually address one called Price Cap regulation by working through another example.

Assume the following:

- **Pn** stands for the new price.
- **Po** stands for the Old or current price.
- **MPR** stands for the Minimum Productivity Factor and is a factor determined by the regulators. This is a factor that if telephone companies exceed, they will be allowed to raise prices. The incentive is to improve efficiency and beat this factor; one can then increase prices and reap greater profit. Of course, it becomes more difficult each year to improve upon the previous year's productivity gains.
- **RPF** stands for the Real Productivity Factor, which is the productivity that the telephone company actually performs.

The following is a simplified Price Cap formula, which is used more to demonstrate the concept of incentive based regulation than actually used in practice: **Pn = Po(1- (MPF - RPF))**

Let us now apply this Price Cap formula to the following problem. It is now the 1990's and a small telephone company and their state PSC have changed to Price Cap Incentive Based Economic Regulation in an effort to make telephone companies within the state more efficient and also less costly for customers. The PSC has set the minimum productivity factor at an aggressive 5% per year. The small telco has worked hard and improved to the point of having a 6% real productivity factor per year. What is the new monthly price that the telco would be able to charge a subscriber if the original price for local dial tone service is $9.33 per subscriber, per month?

Using the formula: **Pn = Po(1- (MPF - RPF))** and substituting for the variables yields **Pn = $9.33 (1 – (.05 - .06)) = $9.42** per subscriber, per month.

As one can determine, there is an incentive for a telephone company to improve their efficiency under price cap incentive based regulation. Incentive based regulation has replaced most rate of return regulation today. Another concept of pricing we'll introduce here may need additional research, study, and explanation in order to fully gain benefit from it. The concept is *Ramsey Pricing*. It's a term that was considered very much during the days when the telephone industry was thought to possibly be a natural monopoly. Since we now try to regulate in such a way that most regulation attempts to emulate pure competition, some may argue that Ramsey Pricing is no longer of any use. However, if one ignores the concept of monopoly for a moment and concentrates on the following definition provided by Business Dictionary.com we may gain additional insight.

Ramsey Pricing is a type of pricing that "argues that if prices are to be increased, it is a good strategy to increase the markup on goods with the most inelastic demand, because consumers or users will buy them anyway. The rule is applicable in taxation too where goods and services are to be taxed but not where all are to be taxed. It was originally proposed by Frank Ramsey and is called the inverse elasticity rule."[59] With this in mind, if one were to need to increase rates or taxes on two specific products that are often purchased in a workplace breakfast/lunch room, Ramsey pricing would likely indicate that it is fine to increase the cost of coffee because people will purchase it regardless of the price increase. Some may even be addicted to coffee. However, it may not be so good to increase the cost of an English muffin because it is more discretionary. Consumers may just find it healthier and more beneficial to do without the English muffin when faced with a price increase.

Last, please recall that telephone economic regulation exists in ⑤⓪ order to protect the public. Most of the time, it is now practiced in such a way as to emulate free trade competition. Sometimes it succeeds. As we progress to a much more competitive telecommunications model, we'll likely observe more free-trade market competition, which, if done correctly and left to actually work as portrayed by the famous "invisible hand" theory, advocates suggest that it will self-regulate.

Summary

Networks are of many different types. IP networking at this juncture is a destructive technology relative to traditional TDM circuit-switched networks and the various laws, revenue streams, structures, and policies that arose during the prominence of the intelligent PSTN. More importantly though, IP networking is a technology that offers great potential for new and innovative applications that will be developed over converged IP networks. However, it will take some time for IP networking to completely overtake and outplace circuit-switched networks and their supporting structures and concepts.

Within any network, addressing is necessary to connect endpoints. Within the traditional PSTN, the North American Numbering Plan and globally the ITU E.164 standard are the addressing schemes of choice. IPv4 and IPv6 are the standard addressing methods for IP networks including VoIP. One approach that facilitates intersystem addressing is ENUM. Addressing will continue to be a basic precept for our networks, but the need for short, simple, and geographically-based addresses within telecommunications will likely be supplanted by standard IP addresses with easy-to- remember names or identifiers.

Toll-free service was born of AT&T in 1967 and its evolution offers an interesting microcosm of technology, market, and policy change over time. Likely, it has seen its peak and may be in the last quarter of its lifecycle.

Unlike many industries, the environment of telecommunications is directly affected in a great way by policy decisions. Often it is not the telecommunications policy itself but the change in policy that drastically effects the environment. Change in policy can cause new opportunity, choice, volatility, and threat. It can change the economics of networks and shift relationships between suppliers, competitors, and customers. There were periods of time in the history of telecommunications when a change in policy had a far greater effect on the industry than technological changes. Consequently, it is important to know the "who, what, why, and how" of telecommunications policy as it directly affects the telecommunications environment. There are also certain concepts like externality, phatic, and some telephone economic regulation and pricing models that one should become familiar with and understand in order to be able to fully appreciate the essence of some portions of our telecommunications environment.

Within the realm of policy and regulation, many of the past policy guidelines were based on the assumption that telecommunications exists in a noncompetitive economic model and one in which there seems to be a limitation or scarcity of communications options and/ or communications paths or channels. Many tend to believe that to be yesterday's model although some forms of it still may exist today. Economists are purists. Most economists would say that in a truly competitive economic model, the market self regulates. Any attempt to modify or adjust within such a model usually results in upsetting the model's equilibrium and should therefore be avoided.

Chapter 2 Study Questions

1. Research "telephone directory" and, after learning about its history and revenue generation, postulate where this concept might evolve to in the future.
2. How might the concepts and notions discussed in this chapter differ for cellular or for overall wireless communications?
3. Using at least two research sources, one of which is Wikipedia, research "universal service" and explain what it is and why it has been so important in the United States.
4. Explain the term "externality" and, using a telecommunications example, explain one telecommunications externality example.

5. Research and explain MPLS/GMPLS.
6. Research and explain IP routing versus circuit switching.
7. If you've never used VoIP, do so. Google "VoIP Providers" and give one a try or just try a global provider like Skype.
8. Visit both the NANP and the ITU online and learn about them. What did you learn? What is Neustar?
9. If you hail from a country other than one covered by the NANP, who/what entity administers your telephone numbering plan?
10. What is the World Telephone Numbering Guide?
11. Visit the FCC Area Code information. Prior to 2000, when numbers were facing exhaustion in a geographical area, the typical remedy was an area code split with about half of the geographical area getting a new area code. What is the typical remedy now? Note: There are several NPRM documents on this and also the NANP has information on it. This question is a difficult one to research. Use your investigative skills and you will be rewarded. Note where you found the answer.
12. Search for and read the U.S. Bill of Rights. Search out the 1st, 5th. and 14th Amendments. State them.
13. Visit your state PSC or your country of origin's regulatory body and provide one interesting fact as well as the URL.
14. What is the name of a major international telecommunications policy and standards body?
15. In a truly competitive telecommunications model, in which no firm has monopoly or market dominance, may we expect there to be no regulation other than traditional antitrust rules?
16. From your technology courses, what is the current status and prospect relative to the actual implementation of IPV6?
17. Research and discuss Ramsey Pricing and provide a current example of its possible use.

Notes to Chapter 2

22 Donna Bastein, John Clark, and Geraldine Weber, "AIN: A Smarter Platform for Service," in *Bellcore Exchange* (1993).

23 Wikipedia, "Bellcore," Wikipedia, http://en.wikipedia.org/wiki/Bellcore.

24 Richard B. Robrock, "AIN and Beyond — Putting a Vision to Work," in *Bellcore Exchange* (1995).

25 Ibid.

26 Harry Newton, *Newton's Telecom Dictionary*, Wikipedia, "Internet," Wikipedia, http://en.wikipedia.org/wiki/Internet.

27 AT&T Archive, *Events in Telecommunications History*, 13.

28 Robert J McAleese, "Conserving Area Codes," in *Bellcore Exchange* (1991).

29 James N Deak, "North American Numbering Plan, Numbering Plan Area Codes," in *Bellcore Exchange* (1996).

30 Ibid.

31 Roger Freeman, *Telecommunications System Engineering* (New York: John Wiley & Son Inc, 1989), 87.

32 "Connecticut Research Report on Competitive Telecommunications," (Dr. Richard G. Tomlinson, 1995). 6.

33 Mike Heller, "Phone Numbers on the Move," *Telephony* (1995), 46-50.

34 Harry Newton, *Newton's Telecom Dictionary*, Wikipedia, "ENUM," Wikipedia, http://en.wikipedia.org/wiki/Electronic_Numbering.

35 "ENUM," http://epic.org/privacy/enum/.

36 Wikipedia, "Telephone Switchboard," Wikipedia, http://en.wikipedia.org/wiki/Telephone_switchboard.

37 Ibid.

38 Gary Morgenstern, "800 Number White Paper from AT&T" (AT&T News Release, 1992).

39 Newton, *Newton's Telecom Dictionary*. (CCIS 2006), 207.

40 Francis Duffy, Maureen Fiorelli, and Michael Wade, "Putting 800-Number Portability in Place," in *Bellcore Exchange* (1993). 8-13.

41 Judge Harold H. Greene, "United States V. American Telephone and Telegraph Company, 552 F. Supp 131," D.C. Circuit Court (1982).

42 Victor J. Toth, "Progress Report: 800 Database Access," in *Business Communications Review* (1992), 51.

43 Ibid. 51-54.

44 Ibid. 51.

45 Ibid. 52.

46 Robert A Gamble, "Number Portability — What Happened," in *Business Communications Review* (1993).

47 Ibid.

48 Duffy, Fiorelli, and Wade, "Putting 800-Number Portability in Place."

49 Ibid.

50 Morgenstern, "800 Number White Paper from AT&T" (1992).

51 *Standard and Poor's Industry Survey* (1994).

52 Gamble, "Number Portability — What Happened."

53 Victor J. Toth, "Preparing for a New Universe of Toll-Free Numbers," in *Business Communications Review* (1995).

54 Kevin Maney, "A Glance into the Crystal Ball Hints at a Future without 800-Numbers," in *USA Today* (2006).

55 FCC, "FCC Rule Making," FCC, http://www.fcc.gov/rules.html.

56 "Merriam-Webster Online Dictionary," http://www.merriam-webster.com/dictionary/externality.

57 Ibid.

58 Harry Newton, *Newton's Telecom Dictionary.*

59 "Ramsey Pricing BusinessDictionary.com," WebFinance Inc., http://www.businessdictionary.com/definition/Ramsey-pricing.html.

CHAPTER 3

Divestiture — A Momentous Phenomenon

At a Washington D.C. press conference on January 8, 1982, Assistant Attorney General for Antitrust in the U.S. Department of Justice, William Baxter, and AT&T CEO, Charles Brown, announced a mutual agreement to terminate the DOJ antitrust suit that it filed in 1974–*U.S. vs. AT&T*. This announcement ushered in the breakup of the Bell System a.k.a. the Modified Final Judgment (MFJ) or just plain divestiture.

AT&T agreed to divest itself of all wholly owned Bell operating companies' exchange operations and thereafter would be free of constraints of the 1956 Consent Decree, and at the time Brown thought that AT&T would be free of most other restrictions and regulation. AT&T had been prevented from entering the computer business but would now be allowed into it. While the agreement did eliminate the lawsuit, in fact, "the settlement did not eliminate or even modify a single regulation that applied to AT&T or the operating companies. State and federal regulators possessed precisely the same statutory power over the industry as before…"[61] This was, in part, due to the fact that the negotiated settlement was made only with the DOJ. The existing laws and regulatory bodies at the FCC and State Public Utility Commissions still maintained all of their power and processes.

One of the major principles of the settlement was to separate the monopoly exchange service from the slightly more competitive long-distance and competitive- equipment offerings. This is sometimes referred to as the separation of the *"power and incentive to discriminate."* It was thought that being close to customers provided power to influence them, while long-distance telephone service was the lucrative portion of telephone service. Jurisdiction over long distance was a very valuable incentive to making money. Without this separation, it was felt that true

Assistant Attorney General William Baxter and AT&T Chairman and CEO Charles L. Brown shake hands after announcing the agreement to breakup the Bell System in settlement of an antitrust suit. Courtesy of AT&T Archives and History Center.[60]

competition could not develop. Some major provisions of the MFJ were:[62]

- The twenty-two Bell Operating Companies were divested from AT&T and could continue to offer local exchange service and local and intraLATA[63] long-distance service. They were combined into seven Regional Bell Operating Companies (RBOCs).
- Yellow Pages, the highly profitable directory advertising service, was allocated to the RBOCs.
- RBOCs were permitted to provide but not manufacture Customer Provided Equipment.
- AT&T kept Western Electric and Bell Labs.
- The RBOCs were given exclusive use of the "Bell" name and logo. Within the new AT&T, only Bell Labs could retain the use of the "Bell" name.
- AT&T was forbidden to enter the electronic publishing market for seven years.
- RBOCs also could not enter the electronic publishing market.
- The RBOCs could petition the court to waive the line-of-business restrictions if they could show that competition in those markets would not be harmed and that such businesses would constitute no more than 10% of the RBOC's net revenues.

The Modified Final Judgment announced on January 8, 1982, took effect on January 1, 1984, and was a consent decree or negotiated agreement between the DOJ and AT&T. It ended an antitrust suit that was filed in 1974 against AT&T. This broke up the Bell System and created seven Regional Bell Operating Companies that offered basic telephone exchange service and local calling, but not long distance, which was the domain of AT&T and other interexchange carriers like MCI and Sprint.

Telecommunications within the United States in 1982 was arguably the best in the world. Neither the Internet nor wireless communications had yet taken off, but the United States had successfully implemented universal service and had excellent channels of communications that in many ways acted as an "enabler" or added "edge" for business and commerce to flourish, both within the United States and internationally. From today's vantage point, when even most of the lesser-developed countries of the world have excellent wireless telephone service, it is interesting to note that the United States no longer enjoys a real communications advantage over the rest of the world. The playing field relative to the role that communications plays in the world trade game is now flat. In 1982 though, the United States was far and away the leader in telecommunications.

Why, then, break up the provider of this world-class telecommunications network? Had Bell Labs failed to provide the technology of the future? Had Vail's concept of universal service and a one- source provider failed to keep the United States in the forefront of excellent telecommunications? Was there a world telecommunications model better than that of the United States that was proven and one that the United States should strive to emulate? Why did AT&T agree to this settlement? These are questions worth asking and, hopefully, you can't wait to understand the answers to them.

Technology Effects

Throughout these years, technology innovation was great. Bell Labs was *the* preeminent research facility in the world. Others were also innovating and the world was quickly changing. Here is a brief list of some important and relevant activities of the time:

- Bell Telephone Laboratories' invention of the transistor in December 1947 and subsequent development and application of it to the field of telecommunications and electronics in general.

- The development of modern television.
- The development of cable TV.
- Satellite technology and its use in communications.
- Further development of computers and the invention of the personal computer (PC).
- The Bell Labs invention of light wave communications and the laser.
- The development of large scale integration (LSI) and very LSI (VLSI), which enables microelectronics components to combine many hundreds of transistors on an integrated circuit and thereby maximize processing power while minimizing physical size.
- The creation and evolution of the Internet by the U.S. Department of Defense's Advanced Research Projects Agency, the National Science Foundation, and others.
- The creation of cellular telephone service by the Bell System.
- Microsoft's introduction of MS-DOS and Windows.

The Bell System made major contributions to innovation and greatly helped the world advance. However, it is somewhat ironic that technology and innovation did not help the Bell System as much as it may have hurt it. Because the 1956 Consent Decree provided for the disclosure and free use (without patent right protection) of its innovations, technology with proprietary rights granted under the patent process no longer provided technological supremacy. In fact, advanced technology actually made it easier for competitors to invest in new networks and systems to more easily compete with the Bell System. The technology "lock" that first Bell used at the beginning with his telephone patent and then Theodore Vail used relative to the loading coil and long distance to provide a technological edge over its competitors was no longer in existence.

Technology as a delta difference when compared to its competitors was a moot point. An interesting example of this situation actually played out after divestiture. Sprint built the nation's first all-digital fiber optic network, completing it around 1986, and eventually created the famous "pin drop" commercial of quality and voice clarity compared to AT&T's many times larger but mixed copper and fiber network. Bell

Labs invented fiber optic technology and while AT&T had a great network, it didn't seem to really see the need to ramp up the installation of fiber since its copper lines could adequately carry the demand and were well within quality specifications. That is, AT&T didn't understand the need until Sprint's very effective advertisements, including one demonstrating that a Sprint customer could even hear a pin drop via their all-fiber-optic telecommunications network. AT&T seemed to manage much like the conservative family who purchases an automobile and maintains it well. New features and innovations arrive with each year's model, but they keep their car for eight to ten years to maximize their return. AT&T was stuck in that mode of thinking for a long time. Their expertise with depreciation tables and getting the most out of sunken capital did not yield a competitive corporate model in the mode of a speed boat. Quite the opposite, they were the large battleship that could not turn on a dime, and there were speedboats encroaching on all sides of their territory.

Policy Effects

Throughout most of the nineteenth and twentieth centuries, public opinion and governmental actions tell us that large business and monopolistic control over telecommunications, gas, electric, and other industries like the railroads are good and necessary in order to provide economies of scale and high-quality services at modest and seemingly fair prices. However, as the 80's approached, it was apparent that the pendulum was shifting to a more open and competitive environment for these and possibly all industries. The basic principles that governed the industry in the 20s, 40s and 60s were being challenged so much more as the 1980s unfolded.

Early in the twentieth century, Theodore Vail's concept of a natural monopoly, accommodation and cooperation with so-called "expert" commissions of government and universal service, end-to-end service with no "foreign" attachments to the network, was all considered by many to be paramount to the end of having an excellent, ubiquitous telephone network in the United States and arguably the best in the world. These concepts were epitomized by his 1908 slogan: "One Policy, One System, Universal Service." It was the touchstone of all legal, regulatory, technical, and management decisions in the industry. "Vail's strategy of regulated monopoly, based on technological dominance and the regulatory partnership, survived over fifty years (1907-1959) without a significant setback,"[64] but the setbacks did come.

Some theories indicate that perhaps the Bell System experienced too much success. By the end of World War II, the great advances that Bell Labs had made began to diffuse to other engineers and firms, and they were facilitated by the free availability of advances due to the 1956 Consent Decree. Specific inventions can be patented, but the theories upon which they are based cannot be, and Bell Labs was in the theory business as well as the innovation and device business. One example is the transistor. It could be and was patented; the theory of solid state physics could not be patented. Others could surely use the theory and challenge the Bell System. It is ironic that "these ... breakthroughs were ... the tools which would be used to dismantle the Bell System itself."[65]

Bell's internal policies of pricing long-distance service (first thought to be somewhat of a luxury) considerably higher than cost in order to help support local exchange services created what economists would likely call an abomination and also a market failure. The Bell System truly believed in its obligation and right to provide service to the public, end-to-end responsibility and system integrity and, therefore, saw little wrong with making basic local service more affordable for the masses by slightly overcharging for long distance service. This was a market failure that would eventually entice others to vie for this lucrative long-distance market due to its inflated price. There may be a lesson to be learned here.

Specific events that began to foretell the move away from a regulated monopoly toward competition first appeared in the equipment side of the business. The Hush-a-Phone case of 1956 was the first chink or failure in the Bell armor, followed by the more substantial Carterfone decision of 1968. These were eventually followed by the FCC registration program for customer-owned telephone equipment that went into effect on October 18, 1977. Under this decision, customers could purchase (likely from a non-Bell provider) and directly connect to the telephone network telephones and other equipment, which was registered with the FCC. Previously, the Bell System required that these connections could only take place if a Bell-provided, protective connecting device was placed between the "foreign" device and the Bell network. Answering machines and alarm systems had been allowed direct connection since June 1976.[66] It's interesting that in July of that same year, Rochester Telephone Corporation became the nation's first telephone company (a non-Bell, a.k.a. independent Telco) to approve customer-owned terminal equipment through a fuse-type protective device, and it also became the first to sell its subscriber equipment and inside wiring to customers.[67] It didn't take a financial genius to quickly realize the tremendous savings that individual residential customers could accrue by purchasing their own phone and eventually their in-home inside wire. Once this realization took place, many consumers became disillusioned with the fact that they had been, in their eyes, overpaying for their telephone for a very long time.

All this helped to shape a new desire within U.S. policymakers, the FCC, state Public Utility Commissions, Congress, the courts, and the American people. This new burning desire was to create a more competitive model that changes in policy might help to bring about. This desire fueled a faster evolution to this new competitive telecommunications model. The breakup of the large, powerful, and dominant Bell System was thought to be a necessary enabler of the now-desired competitive model.

Market Forces

The lack of technological supremacy in an era that was abundant with great technological breakthroughs was coupled with the U.S. policy evolution from one that believed in, and was often based on the concept of, a natural monopoly to a new competitive environment. If these factors weren't enough to threaten and change the Bell System forever, market forces alone *were* enough to cause the change. Within a democracy that began to pride itself more and more on free trade, entrepreneurs sought out lucrative markets to conquer, make money, and become rich. That is precisely what happened.

MCI was the leading entrepreneurial "David" that took on the "Goliath" Bell System and challenged AT&T on the battle fields of U.S. Telecommunications Policy, U.S. Courts, and the competitive marketplace. Jack Goeken lost his franchise to sell radios after a dispute with the manufacturer and moved into the microwave communications business. Early on, he hired William G. McGowan to become its chief executive. McGowan moved the business to the District of Columbia in 1968 and eventually changed the name to just plain MCI. McGowan was a contrarian. While he knew and understood the rules within the industry, they didn't seem beneficial or logical to an entrepreneur like him. He is generally credited with creating the formidable MCI Company that bucked the system and initiated new telecommunications regulatory policy for the United States. Were it not for William G. McGowan, the study of networking and telecommunications technology might have remained the domain of the Bell System and independent non-Bell telephone companies.

Until MCI's breakthroughs, Bell had managed to limit incursions of competitors to what it saw as the periphery of its business: Customer Provided Equipment (CPE) and private line service.[68] However, MCI would eventually drive a stake into the heart of the Bell System–that being long-distance service. MCI started out relatively tame. On August 13, 1969, the FCC approved the application of Microwave Communications, Inc. (MCI) to build a private-line microwave telecommunications system between Chicago and St. Louis. AT&T had objected and this issue was pending for six years until finally approved. While the FCC had intended this approval to be a limited experiment, it instead proved to be the beginning of competitive long distance within the United States.[69] Then on June 3, 1971, "the FCC in its Specialized Common Carrier decision gave general approval to almost 2,000 applications for the construction of private-line microwave telecommunications systems similar to the system it had approved between Chicago and St. Louis two years earlier. Most of the applications were from two applicants–MCI and Datran companies."[70] As a result of these decisions, AT&T's enemy seemed to have penetrated its outer defenses, but Bell still controlled the switched network, which was the heart and soul of long-distance revenue.

Unfortunately for Bell, in 1973 MCI filed a tariff for Execunet service. This service allowed MCI's customers to dial a number in a distant city served by MCI and be connected to that number over MCI's "private lines," even though the customer had not individually leased a private line from MCI. The MCI theory was that the customer could "lease" the line just for the duration of the call. This allowed MCI to have customers share its private lines, thereby making it a switched common carrier that provided long-distance service. From the user's perspective, Execunet was nearly identical to everyday long-distance service that had been the sole domain of AT&T. When AT&T realized this and complained to the FCC in 1975, the FCC directed MCI to stop this practice based on the previous decision that only provided MCI and others with private-line authority via the Specialized Common Carrier decision. However, taking a page from the often used Bell defense, MCI argued that it was already providing the service. After several other volleys back and forth, the D.C. Appeals Court said that the FCC, in its Specialized Common Carrier decision, had not adequately defined the rules and that the Communications Act of 1934 did not grant AT&T a monopoly over long distance and, in any case, could not do so without evidence that such a monopoly was in the public interest. MCI was, therefore, free to offer Execunet unless and until the FCC made such a finding.[71]

These court findings were upheld by the U.S. Supreme Court, but Bell announced that it would not supply the needed local facilities that would allow MCI's customers to access the new service, since the FCC's previous order on this issue was specifically only for private-line service. The FCC supported AT&T's position, but MCI went back to the D.C. court and again the FCC ruling was reversed in favor of MCI.[72] MCI went on to expand into the residential market and experienced a period of 100% annual growth for several years as it took advantage of these new-found opportunities resulting from court/policy decisions. While still small relative to the tremendous size of the Bell System, in effect, it was nibbling away at what heretofore had been seemingly a protected market. MCI's customers and newly built revenue stream had all been Bell's customers and had contributed to Bell's own revenue stream.

There was another more subtle policy issue that took place in the half-dozen years prior to divestiture, which also directly affected the marketplace. This was resale and arbitrage. Bell introduced a service called Wide Area Telephone Service, a.k.a. WATS in the mid-1960s,

presumably in an effort to encourage businesses to initiate new and additional calling patterns and thereby increase their use of the telephone.
A firm could lease a WATS line and call anywhere in the United States
over this line at a fixed monthly cost that yielded a marginal per call
cost approaching $0.00, or nothing, since a firm paid the flat rate re-
gardless of usage. The monthly cost was less than the cost would have
been if calls had been placed via regular long-distance service. This
veiled, large-volume discount, compared to MCI's pricing and that of
other specialized carriers, made Bell's WATS a difficult product to compete against in the large-volume-calling marketplace.

The FCC had been bothered by this WATS product. One of
the keystone principles of FCC telecommunications regulation was
to insure that telecommunications products and services were nondiscriminatory, and it felt that WATS gave businesses a special price for a
service that seemed to be very similar to plain old long-distance service
but packaged for large users. While the FCC was bothered, it left the
WATS tariff in place as an "unlawful but not illegal" service.[73] Investi-
gating this issue and possibly requiring AT&T to eliminate WATS due
to its discriminatory nature was not a savory task for the FCC, since the
outcome of an investigation might alienate large-volume-calling companies. In a flash of true brilliance, the FCC turned this troublesome
investigation and judgment — one that might have been followed by
restrictions — into a market-driven issue. It did not weigh the evidence
and make a judgment, but instead it removed all resale restrictions on
message toll long-distance service (MTS) and WATS and allowed Bell
to file new WATS rates prior to the removal of resale restrictions in 1981.

The economics of a WATS resale were that an outgoing WATS
line cost about $3,000 per month and if it were kept busy for eight hours
per day making long-distance calls, the average cost per minute would
be about $0.20. However, the corresponding cost per minute for regu-
lar long distance to the identical locations would be about $0.33 per
minute. Under the resale umbrella, a reseller could buy Bell's WATS
lines and connect them to a switch and charge slightly below the average long-distance rate–say, $0.28 per minute–while paying Bell $0.20
per minute and thereby make money on the difference, while the end
users would receive about a 15% reduction in their long-distance costs.

This was win-win for everyone except AT&T, since without the change in resale policy, only the very large customers could reap these benefits while smaller companies and residential customers had to pay standard long-distance rates. Thus, an entire industry of resellers sprang up due to these arbitrage opportunities. Today we can envision "virtual" and "online" enterprises that often involve little capital investment. The resale and arbitrage opportunity was the "virtual" company of the 80s. This reselling and arbitrage was particularly attractive because the resellers need only lease WATS from Bell and need not build their own transmission facilities. Even more important was the concept that companies with limited facilities of their own, like MCI, Sprint, and other specialized carriers, no longer needed to restrict their calling footprint. They could now add WATS lines and thereby allow their customers to complete calls anywhere over Bell's ubiquitous system. In a single policy change, the FCC greatly increased the ability of specialized common carriers to compete with Bell. During the development and build out of MCI's network, it could use its major competitor's facilities to improve its own coverage and to act as a backup for its own system. Typically, calls placed over MCI's own facilities would cost it less than over Bell's, so there was still an incentive for MCI to build its own network, but this improved coverage and backup opportunity was great and also something rarely found in competitive environments.

And so it was the pursuit of profit that greatly spurred on the desire for competition in the long-distance arena and seemed to provide the passion required to change the rules of the game and provide opportunity in a heretofore protected Bell marketplace. The early 1980s saw MCI selling a service at off-tariff (lower) pricing to willing businesses and consumers, coupled with policymakers who were willing to change the rules of the game and an era of technological advances that helped to enable the building of telecommunications networks that could be supplemented via reselling Bell's ubiquitous, excellent network.

Summary

During the late 1970s and in the early 1980s through a series of small and at least partially restricted trials and, yes, some errors–the U.S. telecommunications market slowly experimented with allowing firms other than Bell to serve telecommunications markets. No other world tele-communications market was as ubiquitous and efficient and overall as excellent as that of the United States market, and there were few, if any, experiments that allowed new telecommunications providers into foreign telecommunications markets. After all, in most of the modern world, telecommunications was provided by a branch of the govern-ment, not unlike the U.S. Postal Service. These United States trials and experiments occurred at a time when technology improvements were vastly accelerating–pretty much without patent right restrictions. Furthermore, the United States enjoyed almost 100% coverage of tele-communications service, so the original limited concept of universal service was not really in jeopardy. United States businesses and con-sumers observed the various marketplace experiences of the day and liked their results. Contrarians like MCI's McGowan were cheerleaders for competition and champions of the public's right to lower-cost ser-vices. This further pressured the traditional regulatory policymakers–the FCC, state PUCs, and also the Department of Justice.

In summary, change in the perspective of consumers, regula-tors, the Department of Justice, and prospective competitors took place. While one may say that Bell's perspective may not have changed much, the major demonstrated change based on agreeing to the MFJ was that AT&T was now willing to give up part of its heritage in order to get out from under the tremendous burden of regulation and litigation. It had grown weary of the fight and Bell also sought entry into the seemingly lucrative computer marketplace, which it had not been allowed to enter based upon the 1956 Consent Decree.

A personal comment from the author

Intellectually, I know that the U.S. telecommunications environment wouldn't be what it is today without divestiture. Many good things like competition and innovation were sped along by it. In addition, I must say that my own personal work experience at AT&T post divestiture was more varied and interesting as a result of divestiture. Yet, perhaps at other than the intellectual level, there is something, almost a reverence that still gives me a special feeling regarding the great importance of the predivestiture Bell System. I experienced a similar feeling around the turn of the century, when I felt so proud of Bill Gates for successfully circumventing the proposed DOJ breakup of Microsoft. Since that famous decision, Microsoft has pretty much cleaned up its marketing act. Also, the tremendous philanthropic ventures that Gates and his foundation are now performing to help improve the world sometimes makes it difficult for many of us to fathom why the Justice Department could have "dogged" Microsoft for so long and so hard. Had AT&T's Charles Brown followed the hard-nosed approach of his AT&T CEO predecessor John deButts and/or later, the path of Bill Gates–fight rather than compromise–one may wonder what would have occurred. Charles Brown died at the age of 82 on November 12, 2003. An excerpt from his New York Times obituary is interesting. When Mr. Brown agreed to the settlement, he said, "It is time to act, time to put uncertainties behind us, and to begin reshaping the Bell System to match the requirements of a new era." But only a year later, he said in an interview that the breakup of AT&T had been ill advised. "I think the nation in the long run will be sorry it happened," Mr. Brown said. "It wasn't broken and it didn't need fixing. And if you try to fix something that doesn't need fixing, you don't know what's going to happen, especially when it's done in a hurried way." James H. Evans, former chairman of Union Pacific Corporation and a director of AT&T when it was broken up, said at the time that it pained Mr. Brown to settle the case. "He no more wanted to preside over the breakup than Winston Churchill wanted to preside over the liquidation of the British Empire, but Charlie's been perfect for this job" he said. "He was flexible enough to see that it was time."[74] In any case, the Bell System's divestiture of January 1, 1984 was a grand time of promise for all those who sought to compete for a slice of the lucrative long-distance market and also for the millions of American consumers who hoped and expected lower telephone rates. Immediately after divestiture, confusion became the norm for the telecommunications consumer.

End of Chapter 3 Study Questions

1. During the period of time 1970–1988, there were three Computer Inquiry cases that transpired. Research and explain the essence of each.

2. Google and research the history of MCI. Why might one say that its success was based upon a "contrarian" view of telecommunications rules? Was this good or bad? Where or what is MCI now?

3. Who was U.S. District Court Judge Harold Greene and what effect did he have on telecommunications history?

4. Who was U.S. President at the time of divestiture?

Notes to Chapter 3

60 "Milestones in AT&T History," (1982).

61 Faulhaber, *Telecommunications in Turmoil*.

62 Ibid, 98.

63 "'Lata' is a Local Access and Transport Area defined as a result of the Bell Divesti-ture. Switched calls with both endpoints within the LATA (intraLata) are gener-ally the sole responsibility of the local telephone company while calls that cross outside the LATA (interLATA) are passed on to an interexchange carrier." Per: Harry Newton, *Newton's Telecom Dictionary*, 22nd ed., p.533.

64 Faulhaber, 33.

65 H. M. Boettinger, *The Telephone Book*, (New York: Stearn, 1983).

66 AT&T Archive, 119.

67 Ibid, 118.

68 Faulhaber, 67

69 AT&T Archive, 102.

70 Ibid, 106.

71 Faulhaber, 68.

72 Ibid, 69.

73 Ibid, 69.

74 Andrew Ross Sorkin, "Charles Brown, 82, Former AT&T Chief, Dies," in *New York Times*, http://query.nytimes.com/gst/fullpage.html?res=9F04E2D91738F930 A25752C1A9659C8B63.

Postdivestiture through the Beginning of the Internet Age

The year 1984 was truly a time of new beginnings. The seven newly formed Regional Bell Operating Companies (RBOCs) were born of AT&T via the MFJ and AT&T was reborn as a slimmed-down company now unencumbered by its former high-cost, low-profit local operating companies that composed the seven RBOCs. AT&T continued to enjoy its large long-distance revenue stream, but now this market was much more competitive, and AT&T's revenue was destined to decline. In addition, the long-distance interexchange industry was born again into a fairer system called equal access. Both business and residence consumers were faced with new-found opportunities and challenges that, while confusing partly due to the fact that they now had choices, promised to offer better pricing of long-distance services. Would this savings be offset by increases in local telephone service? It was also a time prior to what is often referred to as the Internet Age, which, for the purposes of this book, will be defined as the post-1996 timeframe–a time when the majority of Americans started using browsers like Netscape or Internet Explorer to access the Internet, and a time when online commerce, communities, and communications began to gain traction. This chapter will exam each of these postdivestiture but pre-Internet, coming-of-age issues.

Regional Bell Operating Companies

The Regional Bell Operating Companies, or RBOCs as they were called, were formed from the Bell System's twenty-two wholly or principally-owned operating telephone companies and were, by mutual agreement, organized into seven geographical regions. The MFJ did not specify that there be seven regions, but rather that AT&T and the operating companies decide on the best structure. It turned out to be seven principally geographic regions. The following is a list of the original seven RBOCs and changes that occurred to them over time including up through today. Figure 4.1 provides a nice pictorial of the original and current structure.

The Original RBOCs and Their Evolution

NYNEX, originally composed of New England Telephone & New York Telephone
* NYNEX was acquired by Bell Atlantic in 1996

Bell Atlantic, originally composed of Bell of PA, Diamond State, C&P Telephone, and New Jersey Bell
* Acquired NYNEX in 1996 and GTE in 2000 and changed its name to Verizon

Bell South, originally composed of South Central and Southern Bell
* Bell South was acquired by (the new) AT&T in 2006

Ameritech, originally composed of Illinois Bell, Indiana Bell, Michigan Bell, Ohio Bell, and Wisconsin Telephone
* Ameritech was acquired by SBC in 1999

Southwestern Bell, originally composed of Southwestern Bell
* changed its name to SBC in 1995
* Pacific Telesis was acquired by SBC in 1997
* Ameritech was acquired by SBC in 1999 and AT&T was acquired by SBC in 2005 and SBC changed its name to AT&T
* Bell South was acquired by (the new) AT&T in 2006

U.S. West, originally composed of Northwestern Bell, Mountain States and Pacific Northwest Bell
* U.S. West was acquired by Qwest in 2000

Pacific Telesis, originally composed of Pacific Telephone and Nevada Bell
* Pacific Telesis was acquired by SBC in 1997

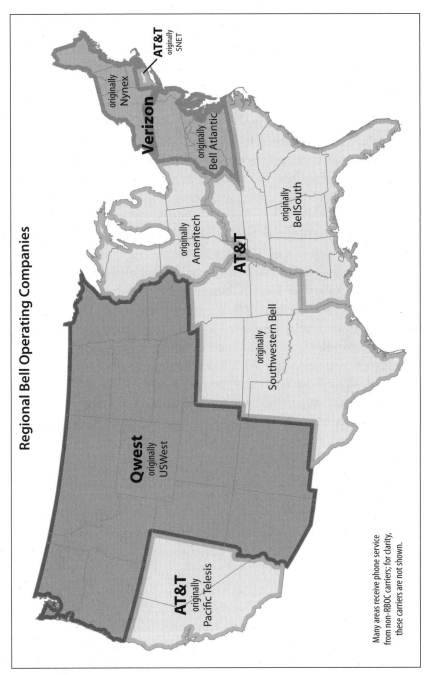

Figure 4.1[75]

Recall that Chapter 3 provides a list of services that these RBOCs and AT&T could and could not provide. Paramount was that the RBOCs could provide local telephone service in the newly established Local Access and Transport Area, a.k.a. LATA, of which there were 196 nationwide. AT&T could not provide service within the LATAs, but could provide interLATA long-distance service. Generally, the RBOCs were forbidden to provide interLATA service, but they were allowed to provide long-distance service outside their home territory where they had little or no facilities and where they were not a dominant player. For example, NYNEX could have provided interLATA long-distance service in California, where it would either have had to build facilities or lease them from Pacific Telesis. This opportunity did not appear to be particularly enticing or lucrative for them.

Of interest may be the fact that AT&T initiated the first commercial cellular mobile service in Chicago in 1983. Some say that Charles Brown and AT&T felt that cellular would "stay with" AT&T, but, in fact, all cellular franchises passed to the RBOCs on January 1, 1984, and so the RBOCs would eventually benefit from this soon-to-be-large, new revenue stream, while AT&T saw their own cellular future opportunity vanish on New Year's Day, January 1, 1984.

Now that we understand the structure of the RBOCs, let us consider their business and mission. Basically, they provided local telephone service to business and residence customers, but were prohibited from providing long-distance service as that was the designated domain of AT&T and the rest of the IXCs and resellers. At the time, local exchange service was not thought to be a highly important and lucrative business and was very capital and labor intensive. However, over time the RBOCs proved that they could make a go of it, and when the Internet took off, the added benefit of millions of customers demanding faster and faster Internet service provided them with a new source of revenue by offering, at first, additional dial-up lines and later dedicated, "always on," broadband access, usually via their ADSL offering. The RBOCs enjoyed two growth areas then–broadband Internet access and cellular communications. Unlike the long-distance interexchange carriers, major disruptive technology changes like today's VoIP actually had an initial positive effect on local exchange providers because the benefits of VoIP work best when enabled by broadband Internet access.

AT&T 1984-1995

On January 1, 1984, the Bell System ceased to exist, and AT&T was reborn without the burdensome local telephone companies and was able to direct all its efforts toward long-distance, manufacturing, and sales of telephone equipment and Bell Labs research and development. After divestiture, Bell Labs was the only place that AT&T could use the Bell name. AT&T had divested two-thirds of its total assets, but kept its then lucrative, but soon to be highly competitive and declining, long-distance revenue stream. While the population was growing, the revenue decline was due to rate reductions and competition.

Why was long-distance now more competitive? The answer is simple. RBOCs were required to treat all IXCs the same and could not provide better service to AT&T than it did to MCI or Sprint or to others. In addition, more FCC policy actions took place that further opened up this market. In 1986 a process called Equal Access Carrier Selection took place. This was a process whereby all long-distance users chose which long-distance carrier they would automatically reach when they dialed 1+ the number. "Some 95 million U.S. telephone subscribers were required to choose a long distance supplier. Those who did not vote–about 30%–were assigned a carrier. AT&T spent $200 million while MCI and GTE Sprint each spent about $75 million in huge publicity campaigns prior to the service election. AT&T remained the preferred supplier."[76] Another series of court and FCC actions allowed all IXCs into the lucrative toll-free service (800 and 8xx) market and eventually enabled this lucrative toll-free market to also become competitive. Again, AT&T's market share and revenues took a dive.

Soon after divestiture, AT&T initiated a 6.4% reduction in long-distance rates, as some of the access-related charges that had been incorporated into long-distance rates began moving to the RBOCs and away from long distance. This was the first of several rate reductions that totaled almost 40% in reduced rates during the six years immediately after divestiture. These rate reductions, coupled with intensive efforts to make AT&T a more market-driven company, helped AT&T to survive, if not exactly thrive, in its newly found competitive environment.

While all of this was going on in the long-distance side of AT&T, its equipment division was working hard in an already competitive arena. Interestingly, another major issue began to be recognized that detracted from AT&T's equipment revenue stream. For years and years, AT&T's equipment division (first named Western Electric, then American Bell, and then AT&T Information Systems since AT&T couldn't use the Bell name anymore except for Bell Labs) supplied/sold just about all the required equipment including central offices to the twenty-two Bell System Operating Companies in the United States. This scenario had been a classic case of vertical integration. Now these twenty-two companies were organized into seven RBOCs that seemed to prefer to not purchase equipment from their former parent and now sometimes rival. AT&T's equipment division sales to operating companies for switching gear significantly decreased. Once again, AT&T saw a major decline in their revenue stream.

With so many declining markets, senior management at AT&T knew that they had to do something extraordinary. One of the motivations for CEO Charles Brown to seek elimination from regulation and agree to the MFJ was that AT&T foresaw some of its markets about to decline but felt that there was a real synergy between AT&T's core business and computers. After all, the largest network in the world was controlled by none other than large central office switches whose control modules were actually high-speed computers. It was Bell Labs scientists Ken Thompson, who created UNIX in 1969, and Dennis Ritchie, who created the C Programming Language in 1973. Now AT&T sorely desired and felt that it needed to enter this hot computer marketplace and make some money. Immediately after divestiture, AT&T was allowed to enter the computer arena, and it did enter it but with unanticipated results. In an attempt to bolster its efforts in this market, AT&T sought out expertise and formed a strategic alliance with Italy's Ing. C. Olivetti & Co. AT&T saw this as a way of merging its voice and data expertise with a proven office machine and computer manufacturer who had a sound European distribution channel. Unfortunately, the strategy never came to fruition, as differences in vision and a lack of communication between the two companies seemed great. Quantitatively, "AT&T

was clearly ineffective in marketing Olivetti's products in the United States, and likewise, Olivetti proved unable to market AT&T's products in Europe. The alliance may have been plagued by a mismatch of expectations, augmented by real cultural differences in their respective markets."[77] After several other misplaced initiatives, AT&T purchased computer manufacturer NCR in 1991 for $7.4 billion and eventually renamed this entity AT&T Global Information Solutions (GIS) in 1994. This AT&T subsidiary had a net loss of over $7.0 billion by year end 1993 and continued to lose money in 1994 and 1995, thereby requiring large infusions of money by the parent company AT&T.[78] It's telling that the October/November 1995 Journal of Financial Economics had an article entitled, "An Analysis of Value Destruction in AT&T's acquisition of NCR."[79]

After trying unsuccessfully to forge a positive revenue stream in the computing marketplace, AT&T changed the name AT&T GIS back to NCR in 1996 and spun it off. AT&T's perceived advantages of vertical integration seemed to be outweighed by costs and other disadvantages. NCR felt a major disadvantage was that many of its potential clients were actually communications services competitors of AT&T and thus reluctant to make purchases from an AT&T subsidiary. This seemed to echo the issue that AT&T's equipment division had posed. In any case, the time that NCR spent as a subsidiary of AT&T was not generally considered to be a positive experience by many employees or residents of its headquarters in Dayton, Ohio. The overall AT&T vision was to show the positive integration of AT&T telecommunications with their NCR based computing power. This was supposed to yield a great synergistic force. Unfortunately for them, this vision never became reality.

Coincident with this foray into the computer marketplace, AT&T initiated efforts to regain a growth engine within one of its core areas of strength. AT&T had invented cellular communications and launched the world's first commercial cellular telephone system in Chicago just prior to divestiture. Not being allowed to keep this gem of potential wealth greatly disappointed the leadership of AT&T. Cellular had been one of the growth engines that it had counted on. AT&T attempted to right this error by announcing the purchase of McCaw Cellular Communications, Inc. in 1993, completing this acquisition in 1994 and naming it AT&T Wireless. McCaw was the largest provider of cellular service in the United States. This was, in effect, the new beginning of AT&T's successful venture into cellular communications. At the time, Bob Allen, AT&T chief executive officer, said, "This merger with Mc-Caw, the technology leader of the wireless industry, is absolutely central to AT&T's networking strategy and key to the company's future-earnings growth. It's a natural fit with our goal of offering customers anytime, anywhere communications. We particularly value McCaw because of the entrepreneurial spirit its people have displayed in making the company the leading provider of wireless services in North America."[80] Craig McCaw, McCaw chief executive officer, said, "AT&T and McCaw are natural allies. We share a common vision of personal communications services that reach people, not places, and our combined strength will enable us to make our vision a reality more quickly."[81]

With so much occurring within AT&T, senior management decided that it would be more efficient and likely lead to more expeditious management action, when necessary, if it separated its major disciplines into separate companies. Also, both its equipment and computer divisions seemed hindered by being a part of AT&T, since its competitors neither desired to purchase telecommunications equipment nor computers from a subsidiary of a major competitor. In the latter half of 1995, AT&T announced that it was voluntarily separating into three separate companies: a services company that included long distance and cellular and would retain the AT&T name; a products and systems company that would be named Lucent Technologies and penetrate sales into the traditional AT&T competitor base; and a computer company that reassumed the NCR name. The cover of the 1995 AT&T annual report was descriptive.

At the time of this "trivestiture," announcement in 1995 and its actual occurrence by the end of 1996, AT&T turned out to be valued at approximately $50 billion, while Lucent was about a $20 billion company.

Divestiture directly and greatly affected the RBOCs and AT&T. Its real purpose was aimed at increasing competition in the long-distance market and thereby lowering business and residential consumer long-distance rates. Divestiture accomplished this goal, and there was only a modest increase in local rates. Anyone with a phone who lived

Figure 4.4.
1995 AT&T Annual Report Cover

through the postdivestiture/pre-Internet period can likely attest to receiving numerous sales calls from the many interexchange carriers and resellers. (This was prior to the National Do Not Call Registry, which came into being in 2003.[82]) Usually these calls would occur at the most inopportune time — for example, at the dinner hour. This experience was a clear indication of tremendous competition in the long-distance arena with the top players being AT&T, MCI, and Sprint, as well as resellers. Many non-Bell independent telephone companies also began to offer their own interexchange long-distance service in order to reap benefits from this lucrative market. A classic example of this is Rochester Telephone in Upstate New York. Prior to 1984 most of its subscribers had Rochester Telephone local service and AT&T long-distance service. After that time, all IXCs were available, but Rochester Telephone got into this market itself and was able to offer excellent one-stop service to its customers and thereby reap benefits from the lucrative long-distance market. Rochester Telephone was a local exchange company, an equipment company, an IXC and also a directory- services-offering company for not only itself but for other independent telephone companies as well. Non-Bell independent telephone companies like it never had MFJ restrictions placed upon them.

As we approach a detailed discussion of the Telecommunications Act of 1996, it's appropriate to mention that Rochester Telephone took the name of Frontier Telephone and offered what amounted to a precursor to much of what eventually became Title I of the Telecom Act of 1996. Frontier called this plan their Open Market Plan and it became effective on January 1, 1995. Key provisions of this plan were total service resale, Unbundled Network Elements (UNEs), and interconnection requirements so that other telephone companies could enter the Frontier geographic territory and compete without building their own network. There was much national interest regarding this plan since it was the only such "Open Market" plan in the nation. It was studied by regulatory agencies, business executives, and Congressional committees. While the proactive Frontier Telephone Open Market Plan was likely a bit more incumbent-friendly than the Act, much of Title I of the Telecommunications Act of 1996 was modeled after Frontier's Open Market Plan. One might ask why Frontier Telephone would open up their protected territory by initiating such a model. The answer lies in the fact that it wanted something from the NY State PSC. Frontier Telephone owned many other small independent telephone companies and sought approval to create an umbrella holding company for Frontier that would optimize rate and policy making and help bring about more efficient management. Likely too, Frontier believed that there really would be a Telecom Act soon so they really had little to lose and might actually be able to better help shape the Telecom Act through their model. Frontier's Open Market Plan was superseded by the Telecommunications Act of 1996.

Summary

This postdivestiture/pre-Internet age was truly a time of new beginnings. Policy change as outlined by the divestiture agreement tore the Bell System into seven RBOCs and AT&T in such a way as to separate the "power and incentive" to discriminate. It was thought that being close to the customer yielded power, while long-distance, unlike today, was a lucrative incentive. Long-distance rates decreased partly because they no longer subsidized local rates and partly because of competition, while local rates that had been subsidized by long-distance rates did rise. This was not enough to significantly offset the nice large decrease in long-distance rates. MCI, Sprint, and other IXCs began to grow and prosper, so it's possible that both consumers and IXCs other than AT&T were "winners" during this time period. The RBOCs managed to survive in their rather protected local exchange marketplace and their cellular service began to grow significantly. Meanwhile, AT&T tried to compete in both long distance and its equipment services divisions where it found difficulties selling into the RBOCs. In effect, these two revenue streams demonstrated significant decreases. AT&T also blundered in its attempts to enter the new computing marketplace and eventually left that market. Cellular was the one AT&T revenue engine that began to hum. If one were to rate the RBOCs as somewhat neutral relative to their successes and failures during this time period, one would have to rate AT&T as demonstrating a decidedly negative set of outcomes and results in their newly sought-after competitive environment.

Considering today's telecommunications environment with much more advanced technologies, including IP networks and VoIP, we might ponder the issue that presented itself to the U.S. Department of Justice back in 1982 and come to a much different conclusion. Today, long-distance calling is not a great revenue incentive for telephone companies. Circuit-switched long-distance customer costs have declined to an area of only several cents per minute due to technology improvements and competition. VoIP further degrades the revenue stream so that a long-distance call might be less than one cent per minute. This is not an incentive that allows a business to make a lot of money. In fact, today one must wonder if there is any possibility of a carrier making money in VoIP other than the peripheral possibilities like advertisement and add-on services. (Yes, it's understood that VoIP equipment vendors can make money and that users can save money with VoIP.) If you're still with this, then perhaps you may be able to easily recognize that the technology advances over the past twenty-five years have perhaps made an environment in which there would be little need to separate the "power and incentive" to discriminate, since the long-distance revenue incentive has vanished. You may also be able to infer that, had VoIP been available in the early 1980s, perhaps the DOJ might not have needed to breakup the Bell System–at least not with the intent of creating competition in the long-distance arena and decreasing the cost of long distance for all consumers. After all, how low can you go?

Another conclusion might be that since VoIP is here today, long distance and cellular are competitive (even if cellular has not yet embraced VoIP); perhaps one need not be so concerned about the size of telephone companies. Later in this book, you'll learn some reasons why size and scope may become very beneficial to the individual companies, their customers, and even the business environment in general. Overall, the MFJ accomplished what the DOJ desired and it would be proper to give it credit for providing public benefit by its actions. It created a new and competitive playing field for long-distance providers and lowered long-distance service costs while allowing enough positive business for the RBOCs to survive. Do you think that the other party to the MFJ agreement–AT&T–got what it desired and bargained for?

Divestiture was about AT&T and the "Baby Bells" that it spawned and the business ventures of each. It also greatly affected their customers including both business and residential consumers. We've discussed AT&T and the "Baby Bells" in detail, but what of the rest of the players in the telecommunications environment? Here is a general but succinct summary of some of the other players and how divestiture may have affected them:

- Non-Bell Independent Telephone Companies were not directly affected by divestiture. However, some entered the lucrative long-distance marketplace and became one-stop shops for telecommunications services and so they benefited.
- Interexchange carriers (IXCs) like MCI and Sprint grew and prospered now that they were provided equal access by the RBOC LECs and could share the long-distance pie with AT&T.
- The difficulties that AT&T's equipment division experienced positively and directly affected other communications equipment manufacturers who began to enjoy a very positive new benefit. They were able to sell their products to the RBOCs.
- More affected by technological advances in networking than the MFJ was the very positive growth of "router" companies like CISCO. This too adversely affected AT&T's equipment division revenue stream.

Generalization seldom adequately describes something as complex as the post divestiture telecommunications environment. Nevertheless, let us indicate here that in general, the weakening of AT&T through divestiture and the introduction of seven independent RBOCs provided an environment that was easier for AT&T's competitors like IXCs and equipment manufacturers to prosper. It also helped to provide more choice for consumers and we know that choice usually yields more competitive prices and cost savings. The policy of breaking up the Bell System and thereby forcing long-distance telephone calling to become competitive, coupled with the marketplace reaction to this policy, worked. Relative to its original goals, divestiture was a success!

Chapter 4 Study Questions

1. Early in this chapter, it mentions that there was confusion in the industry. What was the confusion and why did it occur?

2. Why were RBOCs allowed to offer interexchange long-distance service immediately after divestiture outside of their home territory?

3. Why did the concept of long-distance resale occur?

4. Calling cards came into being during this time period. Why do you think this occurred?

5. When the Bell System was broken up, what happened to the shareholders of AT&T stock?

6. What was American Bell and when was it formed?

7. Research what effect the Computer Inquiries had on the postdivestiture AT&T?

8. In the early 1980s there was no MS Word. In fact, IBM & WANG were the primary word processing providers. Research and learn about this relative to WANG.

Notes to Chapter 4

75   Wikipedia.

76   Jason Manning, "The Eighties Club Ma Bell Breaks Up," http://eightiesclub.tripod.com/id310.htm.

77   Spekman, Robert and Bolon, Meredith, "AT&T & Olivetti: An Analysis of a failed strategic alliance" (1993), 10.

78   Wikipedia, *NCR Statistics*, http://en.wikipedia.org/wiki/NCR_Corporation.

79   Thomas Lys and Linda Vincent, "An Analysis of Value Destruction in AT&T's Acquisition of NCR," in *Journal of Financial Economics, v39* (1995): 353-78.

80   AT&T and McCaw, "People of AT&T Meet McCaw / People of McCaw Meet AT&T," (Basking Ridge, NJ: AT&T, 1994).

81   Ibid.

82   FCC, "Do Not Call Registry," http://www.fcc.gov/cgb/donotcall/.

CHAPTER 5

Telecommunications Act of 1996

After years of discussion, debate, and compromise, the telecommunications bills meandering through the House of Representatives and the U.S. Senate finally converged to a final bill acceptable to both houses of Congress and which the President agreed to approve. On February 8, 1996, in the Library of Congress reading room, President Clinton signed the bill into law. The Telecommunications Act of 1996 was then a reality. To quote an FCC.gov posting in 1996 soon after passage of this law:

(51) The Telecommunications Act of 1996 is the first major overhaul of telecommunications law in almost sixty-two years. The goal of this new law is to let anyone enter any communications business–to let any communications business compete in any market against any other.

The Telecommunications Act of 1996 has the potential to change the way we work, live, and learn. It will affect telephone service–local and long distance, cable programming and other video services, broadcast services and services provided to schools.

(52) The Federal Communications Commissic
role to play in creating fair rules for this nev

The year 1996 was a time of promise and g
in the U.S. telecommunications arena. Inti
like AT&T and MCI anticipated getting i
cations market, which was estimated to b

per year[84]. They actually had to anticipate that because, once the goliath like RBOCs met certain requirements for opening up their local markets for competition, they would be allowed into the still-lucrative, long-distance market and that, too, was worth about $100 billion per year. On the other side of this coin was the RBOCs' concern about losing their protected local exchange market to the many who coveted a share of it. Now local exchange companies must open that market up to any and all competitors. Entrepreneurs of all types eyed the local telecommunications market. Telecommunications attorneys expected considerable work due to the many battles that would be waged regarding the new detailed rules that the FCC would put into place, and consumers hoped that providers might actually fight over them to win their business. There seemed to be a whole new world opening up to many within the telecommunications environment.

Here are two interesting quotes from those days: Patricia Eckert, Former President of the California Public Utilities Commission, said,

> The excitement surrounding the passage of the Telecommunications Act of 1996 is surpassed only by the uncertainty over what it all means. Right now no one knows for sure.[85]

William H. Gaik, Deloitte & Touche National Practice Director for Telecommunications and Electronic Services, said:

> While the pace and scope of implementing the new Act are unpredictable, the direction is clear. As a society, we believe fundamentally that competition brings out the best in people, but we also understand that competition should be fair. While we value a free and efficient marketplace, the system of subsidies and transfer payments cannot, in its current form, lead us there. The road ahead will not be smooth and we will not locate it on any map. The competing interests among the incumbents and new entrants ensure that there will be no shortage of cases before regulatory agencies and courts for many years to come.[86]

we look back at 1996, it is very important to understand the basic
of the United States and how it had significantly changed over
like the time leading up to the passage of the Telecommu-

nications Act of 1934, the United States no longer thought that the provision of services like telecommunications was best served by the monopoly model. Now, the U.S. sought a competitive market model that would yield consumer choice and a variety of levels of service, features, and options. Within the telecommunications environment, the divestiture of the Bell System in the previous decade had demonstrated that a highly competitive marketplace appreciably reduced consumer long-distance calling costs and generally created great opportunities for start-up companies that sought a share of what had been a highly protected and closed market for the Bell System. It is certainly understandable that legislators, regulators, and consumers would desire a similar situation for the local telephone service market. Also, it is very important to understand the things that were not prevalent at the time and, therefore, were not really given much thought during the creation of the Telecom Act of 1996. Probably the most significant thing that was not really addressed in the Act was the so-called Internet Age. While not all that long ago, 1996 was really just the beginning of the Internet Age, and access to the Internet was predominately dial up for the U.S. masses. VoIP, too, was not addressed. Considering the effect that these two major technological advances have had and are continuing to have on telecommunications and networking, it's really hard to believe that they were not addressed in the Act. Other hot topics of today, like the great need for online security and privacy and copyright issues, also were not significantly addressed. WiFi and Wi-Max were not yet invented and wireless communication for the masses was limited to cellular and that was still rather expensive and mostly utilized by businesses and affluent consumers. Network Neutrality was not even on the radar screen.

Act Overview

Let us now turn to the Act. Structurally, the Telecommunications Act was a 128- page law that one should think of as more of an outline of intent. You may obtain your own copy of the original Act by visiting the www.fcc.gov site and searching for Telecommunications Act of 1996. A succinct method for quickly getting to the intent of this law may be found in remarks made by FCC Commissioner Susan Ness before the Cellular Communications Industry Association Special Commissioner's

Forum in Dallas, Texas, March 25, 1996, in which the following was said:

> The beginning is now over. Congress has now provided clear policy guidance.
>
> Here is how the House and Senate conferees described the new law:
>> ... a pro-competitive and deregulatory national policy framework designed to accelerate rapidly private sector deployment of advanced telecommunications and information technologies and services to all Americans by opening all telecommunications markets to competition...
>
> That's a mouthful. But it's not just a lot of words; it's a lot of ideas as well.
>
> Let's parse the language. At the beginning we have the starting point, the guiding philosophy of the legislation: 'A pro-competitive and deregulatory framework.'
>
> And then we have the real objective: 'private sector deployment of advanced telecommunications and information technologies and services to all Americans.'
>
> This is what it's all about—more and better services to consumers.
>
> Between the end of the beginning (the new legislative framework) and the beginning of the end (providing advanced technologies and services to all Americans), we have the means to get to the end: 'by opening all telecommunications markets to competition.'
>
> Your job and mine is to make sure that consumers receive the rich array of benefits Congress intended. Our task is to transform the legislative vision into reality. Competition is the key ingredient.
>
> Success in this endeavor will transform our work and our leisure, will enhance our education and our entertainment, and foster both domestic employment and international competitiveness.[87]

Given this intent, a brief overview would indicate that the Act did open the local and any remaining portions of long-distance markets and cable television service to competition. It removed the restrictions or limitations on the lines of business that the RBOCs, AT&T, and others had as a result of divestiture, and it provided continued protection and expansion of universal service. Overarching principles were to:

- Increase consumer choice.
- Open local markets to competition.
    - Strikes down state and local barriers to competition.
- Provide for cocarrier status to emerging local competitors.
- Protect and even expand universal service.
    - Directs Federal-State Boards to continue to advance universal service
- Specify that all consumers, including low income, rural, insular, and high-cost- area consumers should have access to telecom and information services including long-distance and advanced telecommunications and information services.

The law itself was divided into a series of divisions called "titles."

- Title I—Telecommunications Services outlines the obligations of both new and incumbent local exchange carriers.
- Title II—Broadcast Services addresses all broadcast TV services including satellite services and the process of renewing broadcast licenses. Digital TV is addressed.
- Title III—Cable Services addresses these and also all video programming accessibility.
- Title IV—Regulatory Reform specifies how regulations may be changed and specifies that regulatory forbearance may occur under certain circumstances.
- Title V—Obscenity and Violence addresses obscenity on cable TV and outlines scrambling of channels for nonsubscribers. The "V-chip" is specified here.
- Title VI—Affect on Other Laws outlines the effect of this law on other laws and consent decrees.
- Title VII—Miscellaneous Provisions attempts to cover everything else.

Wikipedia provides us with the following additional, succinct insight into this Act:

> The Act makes a significant distinction between providers of telecommunications services and information services. The term 'telecommunications service' means the offering of tele-communications for a fee directly to the public, or to such classes of users as to be effectively available directly to the public, regardless of the facilities used.' On the other hand, the term 'information service' means the offering of a capa-bility for generating, acquiring, storing, transforming, process-ing, retrieving, utilizing, or making available information via telecommunications, and includes electronic publishing, but does not include any use of any such capability for the man-agement, control, or operation of a telecommunications sys-tem or the management of a telecommunications service. The distinction comes into play when a carrier provides informa-tion services. A carrier providing information services is not a 'telecommunications carrier' under the act. For example, a carrier is not a 'telecommunications carrier' when it is selling broadband Internet access. This distinction becomes particu-larly important because the act enforces specific regulations against 'telecommunications carriers' but not against carriers providing information services. With the convergence of tele-phone, cable, and Internet providers, this distinction has cre-ated much controversy.
>
> The Act both deregulated and created new regulations. Con-gress forced local telephone companies to share their lines with competitors at regulated rates if 'the failure to provide access to such network elements would impair the ability of the tele-communications carrier seeking access to provide the services that it seeks to offer.' (Section 251(3) (2) (B)) This led to the creation of a new group of telephone companies, 'Competitive Local Exchange Carriers' (CLECs) that compete with ('ILECs' or Incumbent Local Exchange Carriers). (This is now more sim-

ply referred to as the requirement of ILECs to sell unbundled network elements aka UNEs to CLECs.)

Most media ownership regulations were eliminated.

Title V of the 1996 Act is the Communications Decency Act, aimed at regulating Internet indecency and obscenity, but was ruled unconstitutional by the U.S. Supreme Court for violating the First Amendment. Portions of Title V remain, including the Good Samaritan Act, which protects ISPs from liability for third party content on their services and legal definitions of the Internet.

The Act codified the concept of universal service and led to creation of the Universal Service Fund and E-rate programs.[88]

Title I

Let us recall that the Act was more like an outline and statement of intent than a detailed law. It gave specific authority and direction to ⟨58⟩ the FCC to develop the details needed to implement it via their due process. Accordingly, the FCC, immediately after passage of the Act, began this enormous undertaking. Over 800 rule-making proceedings were initiated and thousands of pages of Report and Order documents were produced by the FCC in this effort. One could spend much time studying each of the seven titles in the Act and the many FCC Report and Order documents directed to bring the Act's intent to fruition. These are available to you at www.fcc.gov. Although such study would be interesting, we will not pursue that study here. Rather, based in part on Reed Hundt's quote that emphasizes Title I issues below, we will look at a few specific details of Title I and the resulting proceedings. Reed Hundt said:

> Perhaps the most critical economic goal of the new law is that the local exchange market should be opened to competition. This is nearly a hundred-billion dollar business—the biggest monopolized business in our country. There is more than $200 billion in sunk cost undergirding the local telephone company

monopolies. And the service the LECs deliver to 95% of Americans at an obviously affordable rate is critical to our society and economy. Yet, Congress has taken the extraordinary act of opening this market to competition.[89]

The so-called "opening" of the local exchange market was "enabled" by law and pulled along by holding out a "carrot" to Incumbent Local Exchange Companies (ILECs). Please remember this "carrot," as it will be discussed later. Prior to this Act, ILECs invested in and controlled extensive networks and enjoyed a 100% market share of the local telephone market within their service area. This Act changed that and allowed all comers to enter into this market. However, entry wasn't really all that easy, but the Act did take painful steps to somewhat ease this entry. It was conceived that Competitive Local Exchange Companies (CLECs), which were really creations of the Act, would enter the local exchange business via either facilities-based competition or through resale.

- **Facilities-based competition** meant that the CLECs would build their own local exchange network. This was a very expensive and time consuming endeavor. Few CLECs had the money and time to jump into this option. Total service resale or the resale of UNE seemed like a better options to at least get started. Many regulators and others felt that over time, CLECs should and would build their own networks just as long-distance competitors like MCI eventually built their own long-distance networks in the 1980s.

- **Service resale** meant that a CLEC would just buy the entire service from the ILEC at wholesale and resell it to the end user at a retail price and attempt to make a profit on the difference. Earlier, the resale of long-distance service had proven to be a profitable undertaking.

- **Resale and purchase of Unbundled Network Elements (UNEs) was the preferred choice.** This was a system by which the ILEC had to allow CLECs to purchase just about any and all portions of their network at wholesale prices. The CLECs might install their own central office switch then and provide telephone service using some of their own equipment, but likely at least the last mile of network facilities would be purchased from the ILEC.

Please note: In 1996, wireless and cable were not generally considered to be a primary alternate competitive means to provide local exchange telephone service.

It would be an understatement to merely say that there was a bit of a dispute between what the ILECs wished to charge the CLECs vs. what the CLECs were willing to pay. These were wholesale rates and were supposed to be fair to both the ILEC for the use of their network and/or network elements and also fair enough to the CLEC so that they could increase the rate, charge the end user, and make a profit on the difference after their own expenses were considered. Remember that the ILEC had basically invested much into creating their network and now were being required to sell it in whole or in part at wholesale rates to a competitor. The law said that, "a State commission shall determine wholesale rates on the basis of retail rates charged to subscribers for the telecommunications service requested, excluding the portion thereof attributable to any marketing, billing, collection, and other costs that will be avoided by the local exchange carrier."[90] This was the definition for total service resale and was pretty clear. The 1996 Act envisioned total service resale as an interim means of entry into the market by allowing entry without much capital investment. Once a competitor gained a viable market share they would likely build their own network and become a full-fledged facility-based competitor. The economics of total service resale were just not positive to create competition. Try as they did, CLECs found that they couldn't make any money via full service resale. Their own marketing, advertising, billing, and collections expenses combined with their lack of economies of scale and the wholesale payment to the ILEC left them with razor-thin profit margins or in many cases, losses. Total service resellers provided nothing in the way of economic value to add to the telecommunications local service market place and so they exited the market after finally arriving at this realization. One other contributor to this market place failure was that customers had a realization that telephone service purchased through a full service reseller was somewhat of a degraded service compared with service directly from the ILEC. A reseller's customer who had a network problem needed to call the reseller and not the ILEC to report the problem. Then the reseller would open a trouble ticket with the ILEC and the reseller played middleman. New installations and changes also required a two-step process from customer to reseller to ILEC and this resulted in additional time to get anything done. Accordingly, would-be reseller customers had an expectation of a deep discounts when compared to the retail prices obtainable directly from the ILEC and so these CLECs had to keep their retail rates very low.

Considering that full service resale was a failure, regulators and CLECs concentrated on the purchase and resale of Unbundled Network Elements (UNEs) as a method of entry to the local telephone exchange market. UNEs were typically priced low based on a pricing model invented by the FCC in 1996 called Total Element Long Run Incremental Cost or TELRIC. This pricing model assumes that the incumbent just installed a brand new and perfectly designed and sized network of loops, switches, and interoffice facilities using the latest and greatest technology and without any costs arising from embedded or outdated plant. One might call this a Utopian "green field" approach. The UNE model was originally designed to "round out" a CLEC's network. For example, if a CLEC installed a central office switch but needed last mile loops to reach the end-users, they could purchase these last mile loops as UNEs from the ILEC. However, canny CLECs who may have failed at a profitable entry to this market via the full-service resale approach soon realized that they could put lots of UNEs together into a "platform" that came to be known as UNE-P. They would have the equivalent of a total service resale solution with potentially an effective "discount" of as much as 50% off the retail rates compared to something like a 15%-20% discount that had been offered via the full-service resale approach. This allowed aggressive competitors like MCI to undercut ILEC retail rates, make a profit, and succeed in the marketplace. It yielded a reduction in consumer rates that was often enough of an incentive to motivate customers to leave the ILEC and move to the CLEC. In addition it should be noted that UNE prices were (and still are) month-to-month rates from the ILEC to the CLEC so the CLEC bore almost no investment or market risk. They only had to pay the ILEC for the precise number of UNI-Ps that they needed and if a customer departed, they simply canceled the UNE-P package for that customer. This was a "sweet" deal for CLECs.

Incumbent Local Exchange Telephone Companies loudly pro- tested this sweet CLEC deal because they were being forced to subsidize their competitors' profits through a flawed pricing model that caused them to lose money on most UNEs and also forced the ILEC to carry all the investment risk. Why? The reason was so that CLECs could easily enter the local telephone exchange market, become a successful competitor, and steal the ILEC's customers. ILECs didn't feel that this was fair. Over time, the FCC came to realize this as fact and agreed with the incumbents and eliminated UNE-P during a transition period from 2005-2006. By that time, many CLECs had been well fed and succeeded by means of this market failure disgrace. Today, UNEs are still available, but only for their originally designed purpose, which was to "round out" a competitor's network. If a CLEC has no network, it may not purchase UNEs.

Local telephone exchange competition as contemplated by the Telecom Act of 1996 was not successful because the three methods (facilities-based competition, full-service resale, and purchase and resale of UNE) available to a potential new competitor as a means to enter the marketplace were too costly and not profitable. As a result, CLECs abandoned the market and that includes the CLEC activities of the giants like MCI and AT&T. Some ponder if the temporary "sweet" CLEC deal may have improved cash flow enough for some of these IXC and now CLEC competitors to the point of possibly masking the progressive downturn in traditional long distance revenue. Without excellent and detailed measures and metrics, it's often not unusual to lose sight of the obvious when cash flow and overall revenue is good. Success — even if it is fleeting — may become distracting.

Perhaps we should get back to this Title I "carrot" that was held out to ILECs and more specifically to the RBOCs who had been prohibited from offering long-distance service to their customers as a result of the

MFJ of 1984. Long-distance service was still a fairly large and lucrative business in the mid-1990s, and those who engaged in it usually were able to reap handsome profits. Allowing the RBOCs into the lucrative long-distance market was the carrot. This Act effectively nullified the RBOC restriction from providing long-distance service once certain, so-called checklist items were satisfied. This checklist required:

- Interconnection (for CLECs) at any feasible point with the same quality, reliability, and feature functionality that the ILECs provide to themselves for their direct clients
- Number Portability
- Dialing Parity
- Nondiscriminatory Access to Operator Services, Directory Assistance, and Listings
- Resale
- Reciprocal Compensation
- Access to Right-of-Way (like telephone poles)
- Unbundling of Networks yielding UNEs

We've discussed UNEs and Resale already. At this time, we'll discuss two additional interesting items on this checklist: Dialing Parity and Number Portability.

- **Dialing Parity** is the notion that customers who receive their service from a CLEC should be able to dial in the same fashion as customers who receive their telephone service from an ILEC. Assume for the moment that Ron's ILEC has been the sole provider of telephone service in an area for many years and that its customers can make local calls by dialing seven digits and that they can make long-distance calls out of their current area code by dialing a 1 plus the NPA and seven-digit Central Office and Line Number. Now let us assume that a CLEC comes to town to compete for telephone service. It purchases UNE from Ron's ILEC and begins to offer service. If Ron's ILEC requires the CLEC customers to dial thirty-nine digits to make a long-distance call, likely their customers wouldn't like that and might think twice before they sign up with a competitive LEC instead of Ron's ILEC. This type of discrimination is specifically forbidden under the dialing-parity section of the checklist. An ILEC must provide UNEs and resell service to a CLEC in such a way that dialing parity is maintained regardless of whether the customer uses the ILEC or the CLEC for telephone service. In effect, this was an antidiscrimination requirement.

- **Number Portability** is a notion aimed directly toward enabling competition and actually making it more likely to occur. There are four types of number portability under the number portability umbrella. The first type was Toll-Free, a.k.a. 800 Number Portability and you read about it in Chapter 2. The Act defines number portability as, "the ability of users of telecommunications services to retain, at the same location, existing telecommunications numbers without impairment of quality, reliability, or convenience when switching from one telecommunications carrier to another."91 This definition, called Service Provider Number Portability, is used to define local number portability between local exchange service providers and is required under the Telecommunications Act of 1996. It was further specified in the First Report and Order in 1996 by the FCC and it became law! As long as we are discussing number portability, the other two types are Service Portability and Location Portability. Service Portability refers to the portability that arises when a unique service is subscribed to such as ISDN or cellular. If the subscriber wishes to change services, Service Portability would allow this to occur without changing telephone numbers. For instance, Service Portability would allow the subscriber to port their cellular number to Plain Old Telephone Service (POTS) or POTS to cellular. Service Portability for a landline to cellular and vice versa became law in November 2003. The last type of number portability has not yet come to fruition. Location Portability refers to the ability to change neighborhood, nearby community, or state without changing telephone numbers. In effect, should you live much of your life in one area and decide to retire to another location hundreds or more miles away but within North America, Location Portability would allow you to relocate and to take your entire and familiar ten-digit telephone number with you. All your friends and relatives would be able to continue to reach you by dialing your same telephone number. What a wonderful world this would be!

Why was number portability thought to be required? Prior to the Act being passed, major would-be CLEC players like MCI, MFS, and AT&T made significant information available to the FCC and to Congress and entered this information into the official transcripts and records. This information indicated that without number portability, there would be a very real barrier to local exchange market entry. One well-quoted study was commissioned by MCI and completed by the Gallup Poll. It showed that without number portability, both business and consumer telephone subscribers were unlikely to switch basic telephone exchange suppliers. Please see the following chart, which visually shows that requiring a number change is a true deterrent to changing one's local exchange provider. The significance lies in the difference between the percent of consumers and businesses that are likely to change providers with and without a number change.

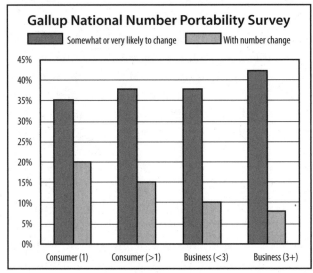

Source: Gallup National Number Portability Survey commissioned by MCI in 1994[92]

A Retrospective View

It has been over a dozen years since the Telecommunications Act of 1996 became law. Many people now wonder if the Telecommunications Act was a good law and if it accomplished its major intent and goals. Anyone who attempts to address this would likely be biased in one way or another. Nevertheless, the following will attempt to address the question. Be aware that it likely will be partially biased based on the author's experiences, vantage point, and political and socioeconomic view of the telecommunications environment and where it fits into the larger sphere of life in the United States.

  Let us start by being reminded that the Act was merely a 128-page law that was an outline of intent. It directed the FCC to fill in the blanks via their rule-making processes and thereby provide the details that would "enable" the intent and, potentially, bring the intent to fruition. Also, we should be reminded that those major issues, like the Internet Age, WiFi, and Wi-Max, and other current-age breakthroughs that occurred after 1995, were not intended to be addressed in the Act and, therefore, were not addressed well, if at all. With that as a springboard, one cannot overemphasize the importance to the change in mindset of all stakeholders that the intent of the Act provided. Recall from the previous Susan Ness quote that the intent of this act was to provide

> . . .a pro-competitive and deregulatory national policy framework designed to accelerate rapidly private-sector deployment of advanced telecommunications and information technologies and services to all Americans by opening all telecommunications markets to competition.[93]

It absolutely did enable this mindset. So that is a tremendous plus for this Act. However, Reed Hundt said that, "Perhaps the most critical economic goal of the new law is that the local exchange market should be opened to competition." This, too, did open to competition, but it was not effective. While the Act had many, many provisions, and the FCC further developed even more provisions that would help "enable" local competition, it really didn't happen. The enterprise (a.k.a. business) markets had been competitive for years, but the average homeowner can choose from about as many providers of traditional local exchange telephone service today as he or she could prior to 1996. The reason is that, while the Act opened up the local telephone market to competition and provided many provisions to help it along, such as resale, UNEs, and number portability, it did not guarantee a business case that would allow a CLEC to grow and prosper–especially in the residential marketplace. CLECs that targeted the residential marketplace typically did not grow and prosper. Goliath would-be CLECs like MCI and AT&T soon abandoned this market due to a very poor ROI. Please remember that the "four-legged stool" discussed in Chapter 1 explains that the telecommunications environment is influenced and supported by technology, policy, market forces, and security. Market forces were much different post 1996 vs. 1984, and local exchange service was much more costly to provide than was long-distance service. Further, the so-called "carrot" (long-distance service) that was intended to motivate RBOCs to cooperate with CLECs, their would-be competitors, all of a sudden had shrunk in value by the turn of the century so that the "carrot" began to look more like an irrelevant pea to them.

Even though there is not much local competition relative to the typical local exchange telephone service in the United States today, some believe that the Act was a tremendous success in that it changed the mindset of all stakeholders. The economics of the marketplace also worked in a very fine manner by driving out potential businesses that were not destined to survive. While all this was going on, technology was changing so as to provide much lower-cost cellular service and thereby become a true, viable option to wired-voice communications, not only when one is mobile but also when in the home. VoIP, too, is on a trajectory to someday become the norm. So, while telecommunications public policy was an enabler and did change the mindset, market forces hindered local telephone competition as envisioned in 1995, but technology advances are making it so that perhaps it just doesn't matter. As it relates to what Reed Hundt called the most significant economic goal of the Act, this goal did not come to fruition as a direct result of the Act because market forces held back advances in competition. However, great advances in technology, including: cellular, WiFi and Wi-Max, VoIP, and, the Internet are all helping to bring about the new world of competition. We have arrived at a place that includes competition, but it is a much different competition than originally envisioned by this Act. Overall, we are meeting the intent and goal of the Act, and Title I services are getting more competitive, but not merely due to changes in policy as a result of the Act.

There was much more to the Act than the overarching goal of creating choice and competition in all telecommunications markets and Title I issues. There also is no shortage of individuals and organizations with specific critiques. Here are some of the more vocal issues taken from a Common Cause[94] publication:

- The Act lifted the limit on how many radio stations one company could own. The cap had been set at forty stations. Radio giants like Clear Channel, with more than 1,200 stations, now exist. It led to a substantial drop in the number of minority station owners, homogenization of play lists, and less local news.
- It lifted from twelve the number of local TV stations any one corporation could own and expanded the limit on audience reach. One company had been allowed to own stations that reached up to a quarter of U.S. TV households. The Act raised that national cap to 35%. These changes spurred huge media mergers and greatly increased media concentration. Together, just five companies control 75% of all prime-time viewing.
- The Act deregulated cable rates. They've skyrocketed, increasing over 50%.
- The Act permitted the FCC to ease cable-broadcast cross-ownership rules. Ninety percent of the top fifty cable stations are owned by the same parent companies that own the broadcast networks, challenging the notion that cable is any real source of competition.
- The Act gave broadcasters, for free, valuable digital TV licenses that could have brought in up to $70 billion to the federal treasury if they had been auctioned off. These are to be used for digital broadcast.
- The Act reduced broadcasters' accountability to the public by extending the term of a broadcast license from five to eight years and made it more difficult for citizens to challenge those license renewals.

Most of these Common Cause issues in some way revolve around the notion that we are intellectually what we hear, read, see, and learn about, just as much as our body is physically based upon what we eat. So have a healthy and diverse diet and also live in an environment that is toxin free. A free and open diversity of ideas, ideals, and news is a cherished concept that Common Cause definitely champions. It is for you — the reader — to discern if these concerns have hurt the American public. However, there is one concern that must be addressed here and that is the license situation.

While it is true that the FCC provided existing broadcasters with free digital TV licenses, in return it took back the old analog spectrum and auctioned it off to the financial benefit of the nation. In some other parts of the world, specifically in Europe, the existing broadcasters were allowed to keep the old analog spectrum and sell it themselves for profit. This is sometimes referred to as the "digital dividend." From these actions, one can certainly argue and have differing viewpoints regarding digital TV licenses. Yet, it is clear that the FCC has attempted to be a good steward of this valuable U.S. spectrum resource by their spectrum auctions and redistribution processes.

Summary

The Telecommunications Act of 1996 was a sweeping revision to the previous Communications Act of 1934. Anticipation was great leading up to its passage and there was almost a sense of euphoria in the air as many entrepreneurs and potential entrants to the perceived lucrative local exchange marketplace envisioned great opportunity. Time would reveal that this euphoric feeling was, in many instances, exaggerated and misplaced.

Much had changed during the sixty-two years since the Communications Act of 1934. Perhaps the paramount change was that the U.S. public no longer believed in a market that allowed a protected, single-source provider of telecommunications services. The competitive model had become king. It is safe to say that this public opinion imbedded in public policy drove this law. The House and Senate conferees described the new law as follows:

... a procompetitive and deregulatory national policy framework designed to accelerate rapidly private sector deployment of advanced telecommunications and information technologies and services to all Americans by opening all telecommunications markets to competition...

Structurally, the Act was more of an outline of intent than a detailed law. It directed the FCC and in some cases State PUCs to provide the necessary details to bring its intent to fruition.

This law and the details provided by the FCC did promulgate the notion that all telecommunications markets must be open to competition by all. However, the economics of providing traditional local exchange telephone service made it difficult, if not impossible, for competitors to succeed in its provision to residential consumers. Business consumers already were experiencing nice competition. Because of this, some might then say that the Act was not a success. It is true that the Act did not bring about the anticipated tremendous competition for local telephone exchange service that was envisioned–especially after observing that during the previous decade, divestiture directly brought about competition in the long-distance marketplace. Another leg of the so-called four-legged stool came into play and that is the Technology leg. The introduction of advanced technologies like IP, ADSL, fiber optics, and VoIP allowed cable and others to enter the traditional voice application marketplace and helped to bring about competition. Cellular advanced and cellular costs went down so it too became a viable option in lieu of basic traditional wired voice telephone service. The Telecommunications Act of 1996 succeeded in bringing about a change in mindset and what the policy of the Act didn't accomplish relative to Title I services was eventually accomplished, and more, by the technological advances of the Internet, VoIP, and cellular communications.

Chapter 5 Study Questions

1. Visit the Telecommunications Act of 1996 actual document found on the FCC Web site. Check it out! Visit the Act in Wikipedia. Prepare at least one interesting observation or question that you have discovered or developed that is not covered in this chapter.

2. Was there any "carrot" for ILECs that were not also RBOCs?

3. What caused the significant decrease in value of this "carrot," a.k.a. long-distance revenue?

4. Postulate whether the Act's "carrot" theory use was too late due to the falling value of long-distance revenue.

5. Is there much value in consumer long-distance revenue for carriers today?

6. Were there any "losers" as a result of ILEC RBOCs not becoming aggressive wholesalers for their UNEs?

7. Is number portability occurring in other countries of the world? Give three specific examples.

8. What is the biggest obstacle or complaint relative to number portability?

9. Sometime in the future, all voice communications may occur via VoIP. Are there major databases required to provide number portability via VoIP?

10. Research TELRIC and be prepared to discuss it in class.

11. The term "digital dividend" has multiple meanings and uses. Please describe at least three.

12. Create a list of things that you feel that the Act did not adequately address, but please omit all new technologies and innovations and notions that occurred post-1995. Also exclude the unintended Common Cause issues mentioned in this chapter.

13. Location Portability is depicted as potentially offering a "wonderful world." Please consider this and provide a short discussion relative

to whether or not there might be negative aspects of geographically porting one's telephone number.

14. Consider Titles II through VII of the Telecommunications Act of 1996. Research one of them and write a report describing the major changes that resulted from your chosen Title.

15. Research the power industry in the United States. Compare and contrast the unbundling of electricity delivery and generation with a similar situation within the telecommunications field.

Notes to Chapter 5

83 FCC, "Telecom Act of 1996."

84 Reed Hundt, *You Say You Want a Revolution* (New Haven: Yale University Press, 2000).

85 Deloitte & Touche Consulting Group, "The Telecommunications Act of 1996 — A Comprehensive Overview of the New Law," (1996).

86 Ibid.

87 Susan Ness, "End of the Beginning — Remarks by FCC Commissioner," http://www.fcc.gov/Speeches/Ness/spsn607.html.

88 Wikipedia, "Telecommunications Act of 1996," Wikipedia, http://en.wikipedia.org/wiki/Telecommunications.

89 Reed Hundt, "Words That Matter: Writing Down a Commitment to Competition and Kids" speech, (1996).

90 U.S. Congress, "Telecommunications Act of 1996," (1996).

91 Ibid. Section 3, Definitions.

92 Gallup Poll, "Gallup Poll National Number Portability Survey" (1994).

93 Susan Ness, "End of the Beginning — Remarks by FCC Commissioner," http://www.fcc.gov/Speeches/Ness/spsn607.html.

94 Common Cause Education Fund, "The Fallout from the Telecommunications Act of 1996: Unintended Consequences and Lessons Learned" (2005).

CHAPTER 6

The New Millennium through 2009

The Telecommunications Act of 1996 removed restrictions from the various telecommunications stakeholders and offered the potential to change the way we work, live and learn. This major policy change yielded a time of promise and great anticipation for the telecommunications industry in general as existing and new players anticipated success in markets heretofore restricted to them. To a broader extent, the change to a new century, and even a new millennium, offered tremendous attitude and perspective changes from the old and perhaps now recognized as not-so-great ways of the twentieth century to modern and presumably more positive twenty first century views. This portended an even more vibrant and optimistic view of the future—perhaps one that is almost Utopian or perfect. Can one ever really obtain Utopia? The United States had primed itself for expectations far beyond those realistically obtainable and, in retrospect, the U.S. telecommunications industry was doubly threatened by the positive possibilities resulting from the Telecom Act of 1996, coupled with the overly optimistic and improper expectations of the future. As it turned out, this was a perfect recipe for disaster.

At the same time that general positive feelings offered much optimism, the telecommunications industry saw some segments skyrocket to success while others faltered. The cell phone penetration since 1996 took off and double digit growth was the result. This industry actually didn't falter and continues to be very positive, although it is beginning to taper off. According to a 2008 Harris Interactive survey, approximately 90% of U.S. adults are now cell phone users. This has increased from 77% at the end of 2006. Final 2008 results are anticipated to show growth rate drop to single digits. This suggests that wireless carriers must develop other applications and services if their revenue growth rate is to be maintained. End-user devices like the iPhone will contribute to this development effort.

A most interesting and significant telecommunications segment that was not addressed in the Telecommunications Act of 1996 is the Internet. The turn of the century forecasted growth of the Internet and both user and telecommunications carriers' forecasted tremendous traffic growth related to it. Unfortunately, this incorrectly created the concept of a fast and powerful perpetual growth engine for success. In fact, the new millennium expectation of the doubling of Internet growth nearly every three months was one of the key ingredients for eventual failure within the telecommunications and the dot-com sectors of the U.S. economy. Also, there was too much optimism by those entrepreneurs who sought to enter the local telephone market place as a result of the Telecom Act of 1996. It is as if these would-be successful entrepreneurs considered the tremendous opportunity and successes that resulted from the previous decade's Bell System breakup and opening of the long-distance market to competition and imprudently went after the local telephone service market without adequate market analysis and research. Entrepreneurs just sought to replicate past successes in that last remaining "traditional" frontier within the U.S. telecommunications marketplace. Was that so bad? The failures within telecommunications were further exacerbated by other marketplace failures and finally resulted in a general downturn of the broader U.S. market that became apparent to just about everyone by 2001. These failures were extended when sound, basic financial investment practices began to be ignored and investments began to be made based upon the euphoric expectation of fast growth that wouldn't quit. Former U.S. Federal Reserve Board Chairman Alan Greenspan coined the phrase "irrational exuberance" to depict investments and a stock market that had "gone awry" by overvaluing stocks and other earning assets due to overly optimistic growth and earnings rates. These were sometimes accompanied by imprudent and/or illegal investment practices. Perhaps there is an analogy between the market failures just mentioned and those of the 2007 and 2008 U.S. mortgage crisis and broader economic downturn and recession? Will we ever learn? In any event, the beginning of the new millennium roared into being and demonstrated a disastrous start that may best be depicted by the following NASDAQ stock closing chart.

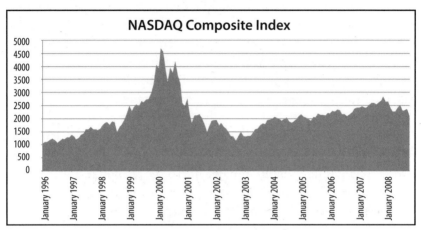

The technology-heavy NASDAQ Composite index peaked at 5,048 in March 2000, reflecting the high point of the dot-com bubble. Source: NASDAQ

There is no doubt that this new millennium was ushered in by an irrational and over- exuberant promise of tomorrow. Further, the telecommunications sector fostered incorrect and overreaching expectations of basic growth rates for the Internet and also for the potential ROI within the newly opened-to-competition local telephone exchange marketplace. The beginning of the twenty-first century saw "communications business made up of companies providing everything from phones to computer networks to routers and switches laid low by the worst collapse to hit a U.S. industry since the Great Depression."[95] When originally written, that statement was true. However, the collapse of the U.S. mortgage and financial markets in 2008 and the corresponding recession will almost assuredly and unfortunately surpass the early part of this decade's telecommunications market collapse. Returning our focus to the beginning of this twenty-first century, let us recall that it was "with breathtaking speed and little advance warning that high-flying companies like Global Crossing Ltd. and WorldCom Inc., which had loaded up on debt to build out fiber-optic networks and buy up companies in anticipation of a never-ending e-commerce boom, collapsed into bankruptcy. Giants such as AT&T were ripped apart as they scrambled to recover from free-falling sales and profits. Hundreds of thousands of workers lost their jobs and prices of some inflated stocks–boasting a price-to-earnings ratio that topped 400 in the most extreme cases–tumbled 95% or more. Investors saw some $2 trillion of market value vanish in a little more than two years–twice the damage caused by the parallel bursting of the Internet bubble. Amid the wreckage, some predicted it could take a decade or more before the industry would climb back and fill all those empty pipes that starry-eyed executives had buried beneath the earth and oceans."[96] Thankfully, recovery has occurred although it's certainly been dampened by the 2008 and beyond recession. This chapter will selectively address some of the issues, concepts, subsectors, and companies that played a major role or were significantly affected during this era. End of chapter questions and guided self-discovery will extend your knowledge beyond the chapter selections and coverage.

The burst of the dot-com bubble and the collapse of the U.S. Telecommunications Industry were really two distinct and different phenomena, although the bubble burst certainly contributed to the collapse of the telecom industry. Historically, an economic boom followed by a bust is a term that's used to describe the cycle of economic upswings and downswings in the economy. There is no shortage of economic theories regarding boom-to-bust economic cycles and their causes and possible prevention measures. Keynesian economic theory (sometimes associated with tax and spend theories) gained favor during and after the Great Depression, while Neoclassical economic theory may be best thought of as "Reaganomics" where less big government taxation and spending is thought to allow markets to work. It also implies that excessive government intervention can only serve to worsen and potentially extend bad economic times. It does, however, encourage the use of the central bank theory and its affect on the money supply and capital investment. There are many more economic theories with much to say about the boom-to-bust cycles. Regardless of economic theory, most consider the cause of such a cycle to typically be based on some irregularity or flaw in the marketplace that causes an extraordinary boom period that eventually is corrected by a bust. Some previous booms of the past include: the railroads in the 1840s, the automobile and radio in the 1920s, transistorized electronics in the 1950s, computer time sharing in the 1960s, and home computers in the 1980s. Moving beyond the Internet boom that led to the 2001 telecom and Internet bust, the U.S. economy experienced a mortgage and banking boom partly fueled by legislation and policies that allowed predatory mortgage lending coupled with general excessive corporate and personal debt to create the bust of 2008. During the 1995–2001 dot-com bubble, an Internet company's growth and survival was thought to depend on expanding its customer base as rapidly as possible even if this produced large annual losses. The term *"Get large or get lost"*[97] seemed to exemplify the wisdom of the day. At the peak of the boom, a promising dot-com company might successfully seek out venture capital funding to start their business and this would often be followed by an initial public offering (IPO) of its stock in order to raise substantially more capital even though the company had never made a profit, or in some cases, had no positive cash flow. A concept called the "burn rate" was the rate at which such a company would run through its capital. The goal, of course, was to gain revenue and become profitable before all funds were exhausted. During the build-up period, public awareness both for financial funding and customer use was important. For instance, during the January 2000 Super Bowl, seventeen dot-com companies paid over $2 million for each thirty-second spot in an effort to gain that sought-after public awareness. Super Bowl 2001 featured only three such extravagant dot-com companies. March 2000 is generally thought to be the time that the dot-com bubble burst. There are many schools of thought as to what exactly caused the bursting of the dot-com bubble. Most, however, would agree that a poor business model coupled with unattainable revenue growth and less-than-prudent financial investments all played a role. Irrational exuberance played a role. Most dot-com failures resulted in the company going out of business–some with and a few without filing for bankruptcy. Much can be learned by examining the past. Accordingly, CNET's Top 10 dot-com flops[98] are worth review. This list is searchable or, as of this writing may be found at: http://www.cnet.com/4520-11136_1-6278387-1.html. The list includes Webvan which was an attempt at being an online grocery business. At its peak, Webvan.com came to be worth $1.2 billion (or $30 per share). In July 2001, it closed and put 2000 employees out of work and left San Francisco's new ballpark with a Webvan cup holder at every seat. This list also includes Pets.com, Kozmo.com, Flooz.com and eToyes.com. The names of some of these flops have been purchased and are being recycled for another try at success.

Earlier it's mentioned that the damage in market value terms within "communications businesses" may have been in the $2 trillion range or about twice the damage directly attributable to the dot-com demise. The question of the day becomes why? First let us recall that the market value of these companies was suspect. Communication- oriented companies tended to be considerably overvalued. The market psyche of "irrational exuberance" had pushed their value far beyond prudent investment valuation, so that when a correction to expectations based upon the realities of the market finally occurred, market value sunk like a lead sinker in a fish bowl. Some homeowners felt this sinking feeling in 2008 when their homes greatly sunk in value. Then too, it is easy to understand that with so many dot-com companies vanishing, the Internet traffic and associated revenue attributable to them would also vanish. Since communications companies had projected tremendous growth rates, their infrastructure and network capital investments were based on these projections, coupled with a return on this invested capital and the revenue generated from the usage, which just did not occur. There was a shortage or loss that appeared to be unavoidable and sustainable. This overleveraged situation with seemingly sustainable forecasted losses pushed many of these communications companies into bankruptcy. "The first big dominos fell in 2001, when broadband providers Winstar Communications and 360Networks filed for bankruptcy. Over the next three years, 655 telecom companies with a combined $749 billion in assets, filed for bankruptcy, according to BankruptcyData.com. On July 21, 2001, after an accounting scandal revealed billions of dollars of overstated profits, WorldCom Inc., the giant that embodied the boom era's promise, filed the largest bankruptcy claim ever (to that date)."[99]

There were other issues that affected the telecommunications environment as well. The technology sector had perfected Voice over IP, or VoIP, communications, which is a considerably more efficient method of delivering voice communications and also a great financial savings to the end user. This method, plus the competition in traditional circuit-switched voice calling, caused that heretofore tremendous voice revenue stream for telecommunications-carrier companies to significantly decrease. Long-distance voice calling had been a really large revenue stream for these companies since the advent of long-distance calling. In fact, the industry was built on it. Now it had very quickly retreated to a revenue stream with a very low profit margin. Consultants must ponder if it is possible to create a profitable business plan via providing long-distance calling to "wired" residential customers. The years immediately after the Telecom Act of 1996 and into the twenty-first century also saw a great influx of Competitive Local Exchange Companies (CLECs) enter the local exchange business in an effort to gain market share and reap rewards that many hoped would parallel the rewards won via competing in the long-distance marketplace soon after divestiture. However, the local exchange marketplace was much different. The cost to market was great since a new CLEC typically had to either invest in their own new network–something that was often cost prohibitive–or they must purchase service or unbundled elements from the ILEC at wholesale prices and resell them to end users at a slightly marked-up retail price and hope to make a profit on the margin. This did not prove to be highly feasible in the residential marketplace for even the largest CLEC entries. Small upstart CLECs, as well as large, well-funded companies like MCI and AT&T, had big plans to win residential market shares in the local exchange business that they had been prohibited from entering until the Telecom Act of 1996. They soon realized that there was little money to be made in this market segment and began to contract their local residential market initiatives. Eventually, they came to realize that this was not a great and profitable market area for them. Further, the adverse effects of little profit from long distance made the likes of AT&T and MCI seek shelter from the market by merging. MCI had merged with WorldCom in 1997. After WorldCom filed bankruptcy in 2002 and eventually emerged from bankruptcy in 2004, it renamed itself MCI. On February 14, 2005, Verizon Communications agreed to acquire MCI for $7.6 billion. AT&T merged with SBC in 2006 and the name of the merged entity was changed to AT&T and is often called the "new AT&T." It is interesting to note that this also was the end of an industry. Interexchange Carriers or companies or IXC as they have come to be known is a U.S. legal and regulatory term for telecommunications companies whose primary offering and revenue stream was long-distance service. The big three IXCs were: AT&T before its merger with SBC, MCI before its absorption by Verizon, and Sprint before it spun off its ILEC services in 2006. In effect, the demise of the IXC business was complete. Long distance became merely an additional product offering of companies that offered much more. Some examples of such companies are: the new AT&T, Verizon, Qwest, and Sprint Nextel.

Before leaving the early part of this decade–a period that turned out to be disastrous for both dot-com and telecommunications companies alike — let us turn our discussion to a former U.S. telecommunications giant with a heritage that traces back to the very beginnings of the telecommunications industry — Lucent Technologies. What happened to Lucent is a telling example of what happened to a fine company that ignored time-tested basic management principles in favor of the fast-paced seemingly "fast-buck" Internet opportunities offered during the boom times of the late-90s and into the next decade. AT&T had been a traditional vertically integrated company. Beginning in the earliest days, it designed and manufactured just about all of the equipment that went into its networks as well as the telephone instruments, copper wire, and eventually fiber optic cable. Bell Labs was the premier research arm of the company and its research came to fruition by being manufactured by AT&T's fully-owned Western Electric manufacturing arm, which was folded into AT&T Technologies when the Bell System was split up in 1984. As mentioned in a previous chapter, when the divisions of AT&T began to sell in a competitive world, some competitors preferred to not purchase from AT&T. Why help a competitor? Consequently, the equipment manufacturing business of AT&T was spun off in 1996 to free it from this market hindrance. Lucent Technologies was born! In the excellent book, *Optical Illusions–Lucent and the Crash of Telecom,* author Lisa Endlich takes the reader through the fast paced but ultimately tumultuous Lucent evolution from an old-line manufacturer to a sizzling hot stock thanks to the emergence of the Internet and the buildup of telecommunications from 1996 through 2000. During those so-called heydays, Lucent President Rich McGinn said the following: "Since 1995, we've taken a $20 billion business and, through 1999, moved it to almost $40 billion, and taken profits from $500 million to almost $4 billion. We describe that as directionally correct." [100] Unfortunately, McGinn would later endure almost universal criticism. Lucent was caught up in the irrational exuberance cycle and became overextended. In the space of two years, it went from the sixth-largest corporation in America and the most widely held stock in the country to being downgraded to a junk-bond credit rating and under investigation by the SEC for its accounting practices as the value of its stock plunged to the price of a cup of coffee. See the following brief financial snapshot of Lucent Stock from its beginning to bust:

Lucent Stock Price		
	Opening Bell: $6.10 the split adjusted price	
1996	High: $10.17	Low: $ 5.70
1997	High: $17.38	Low: $ 8.57
1998	High: $43.61	Low: $14.06
1999	High: $64.48	Low: $36.00
2000	High: $59.36	Low: $ 9.96
2001	High: $17.26	Low: $ 4.09
2002	High: $6.13	Low: $ 0.55

Source: *Optical Illusions: Lucent and the Crash of Telecom*[101]

During the most severe bust years early in the decade, Lucent significantly cut employment by 75%, cut employee and retiree benefits, and generally was a company in crisis that was fighting for its survival. It did survive and improve, but it no longer thrived. One remedy that Lucent prescribed for itself in order to survive was to sell off some of its assets. In April of 2000, Lucent sold its Consumer Products unit to VTech and Consumer Phone Services. In October of the same year, it spun off its Business Systems arm into Avaya, Inc., and in June of 2002 it spun off its microelectronics division into Agere Systems.[102] Around that time, it sought a "white knight" to help it along. Discussions with Marconi, British Telecom, Motorola, and Alcatel took place but without consummating a deal. Then, several years later on April 6, 2006, Alcatel announced it would merge with Lucent Technologies with the headquarters of the combined company placed in Paris, France. Alcatel's acquisition of Lucent was completed on December 1, 2006. On February 8, 2008, Alcatel-Lucent posted a $3.74 billion fourth-quarter loss and canceled the dividend after taking a write-down of more than $3 billion on the U.S. wireless business. Analysts suggested that perhaps Alcatel shareholders had overpaid for Lucent.

Yes, it is clear that during the early part of the first decade of the twenty-first century, the telecommunications industry is *the* model for one of the worst U.S. market collapses since the Great Depression. That is, until the even worse collapse of the housing and financial markets in 2008. While the 2008 collapse and recession is too current to accurately measure the results, from a financial, employment, product output, bankruptcy, and overall business perspective, we can and did measure the telecommunications industry market collapse and it was devastating.

The good news is that the telecom industry has roared back to life and may actually be stronger for going through that collapse and the turmoil that ensued, although it is true that during the 2008-and-beyond recession, few industries are booming, but the telecommunications industry has come back and some might argue that the difficulties experienced in the early part of the decade that left only the strong to survive, have, in fact, helped the telecommunications market sector to better weather the situation that occurred during the latter part of the decade.

Some economists might even suggest that the dot-com and telecom simultaneous collapse helped "kill off" the weak rather quickly and left only the strongest and the well experienced survivors. Credit a return to sound investment fundamentals coupled with a steady rise in appetite for broadband Internet connections as major contributors to telecom's rebirth. Applications like watch-me, watch-my-cat video clips, iPod and other music files, gaming, and Web-inspired services like free voice Internet calling have pushed the adoption of broadband in the United States to beyond the 50% penetration levels. These and a seemingly perpetual introduction of even more innovative applications coupled with interesting and fun-to-experience interfaces seem to be the keys to the ignition of consumer demand. Two additional signs of recovery are the $8.2 billion that Silver Lake Partners and the Texas Pacific Group agreed to pay for networking gear maker Avaya on June 5,

2007 and the June 2008 announcement regarding Verizon's $28 billion purchase of Alltel, the fifth-ranked U.S. cellular carrier with approximately thirteen million subscribers in thirty-four states and primarily thought of as a rural cellular company. This allowed Verizon Wireless to overtake AT&T Wireless as the biggest mobile-phone company in the United States at the time.

During the trough of the telecom bust, there was a glut of broadband communications capacity. In fact, in 2002 about half of the Internet's transmission capacity in the United States was going unused. Even though the pipeline has almost doubled in size since then, the unused portion is down to about 30%, and the price that companies pay for bandwidth in some parts of the United States is on the rise after years of decline.[103] Perhaps one of the best indicators of the telecom revival is that profits for the industry reached an all-time high in 2007 and for the first time surpassed the previous high-water mark of $65 billion set in 1998. Market corrections in part due to the U.S. mortgage crises and gasoline surpassing $4 per gallon in 2008 have had a dampening effect on this and nearly every other sector of the U.S. economy. Perhaps the fact that the telecom sector has implications beyond Wall Street may yet help the overall U.S. economy. Studies show that "a dollar spent on telecom infrastructure produces an outsize impact on the U.S. economy as a whole ... telecom investment plays a vital role in stimulating economic growth and productivity—more so than money spent on roads, electricity, or even education. Communications assets generate massive benefits by slashing the cost of doing business across the economy. A high-speed data network suddenly makes it easier and cheaper for all kinds of workers to place orders, service customers, and drum up new business."[104] So too does the fast, efficient, and ever-improving wireless networks of today. Real estate agents, sales people, law enforcement officers, and Blackberry-toting executives all endorse their mobile access to information. Considering the renewed United States interest in environmental issues, coupled with the need to conserve energy costs, perhaps audio and audio/video teleconferencing and telecommuting may even see a renewed interest now that they more clearly represent economic and environmental benefits.

Specifically addressing some of this forecasted broadband demand are the traditional telephone companies and the cable companies. Unlike in the past, these two industries are not only in direct competition with each other, but will likely blend into only one industrial segment sometime during the next decade or two. Their initiative is based on the so called "triple play" marketing term. In telecommunications, the "triple play" refers to "the provisioning of two broadband services such as high-speed Internet service and television along with one narrow band service such as voice telephone service — all over a single broadband connection."[105] Furthermore, "triple play" relies on the right combined business model with unified customer care and basically a completely unified and integrated service model to deliver the multiple and diverse applications to a single user. At the turn of the century, the cable industry was best technically prepared to offer "triple play" services since their networks already had more bandwidth and could sufficiently handle the delivery of television, broadband Internet service, and the growing interest in VoIP. Their networks were basically fiber and coaxial-cable based. While coax certainly offers more bandwidth than the copper twisted-pair technology present in much of the telephone company's networks, cable companies are in the process of upgrading their networks to be all fiber-optic based. They are actively doing this, but typically they do not forecast and publicize their fiber deployment investments and schedules. Their approach seems to be one of keeping under the radar and then just delivering the goods.

Telephone companies are feverishly trying to catch up in the broadband race. The largest three main line telecom companies in the United States are: AT&T, Verizon, and Qwest. Together they comprise the original seven RBOCs as well as the original AT&T, MCI, World-Com, and GTE. Both Verizon and AT&T have specific, extensive, and well-publicized plans to upgrade their networks to fiber optics. Launched in September of 2005, Verizon has the most extensive plan and will offer it in ten states. It includes spending $23 billion to upgrade to what they call a FiOS (Fiber Optic Services) system that "will be able to deliver data at speeds of 100 megabits per second or more. FiOS has nearly unlimited capacity for video, which is transmitted via a separate, dedicated wavelength. That way, Verizon says, video can't interfere with data transmissions and vice versa. FiOS owes its muscle to Verizon's decision to run fiber lines all the way to customers' homes." [106] AT&T has taken a less bold move to build their "Lightspeed" U-Verse new broadband fiber-based network. AT&T launched "Lightspeed" in June of 2006 at an estimated cost of $4.6 billion in thirteen markets dispersed over five states. "Lightspeed" is envisioned to be able to provide voice, data, and TV services at approximately twenty-five megabits per second. The primary difference between it and Verizon's FiOS is that AT&T's network will not be pure fiber to the home but also will include some copper. In essence, AT&T hedges its bet a little by providing fiber to the suburban neighborhood and then existing copper lines traverse to the homes. When demand dictates, AT&T can upgrade the last link to fiber. Both technologies offer significantly greater bandwidth to the home than traditional copper wire or cable-provided coaxial cable to one's home.

Providers hoping to capitalize on even greater synergies are looking beyond the "triple play" move to something called the "quadruple play." This includes television, high speed Internet, and voice calling, but adds mobile service as well, which of course would be provided via a separate wireless network. Again, the success or failure of such a venture depends more on the quality, reliability, service, and bundle pricing than the technologies deployed. It depends on the value equation presented to the end user. Verizon and AT&T also are the two largest cellular carriers in the United States. Undoubtedly, they hope that their cellular divisions can help support them while they attempt to slow defections of traditional landline telephone service as they prepare the way for their triple- or quadruple-play bundles. One challenge is to overcome their reputations for offering merely "dumb pipes" and prove they can fill their networks with innovative bundles of products and services that strike a chord with customers–all while battling cable operators who already own much of the television subscribers and who are poaching millions of traditional voice-phone customers to their bundled all-in-one voice calling plans that are made possible via VoIP. While all this is happening, new entrants such as Google and Apple are contributing to the mix by offering their sometimes unique perspectives to new and existing applications. With such vast changes occurring, and the blurring of two separate industries eventually into one comprehensive one, it is no wonder that the U.S. regulatory agencies are having great difficulty in fairly and equitably regulating. After all, the regulatory bodies came into existence when telecommunications was telecommunications and television and cable TV was television and cable TV. There was a sort of smokestack structure and mentality among the regulatory bodies. They, too, will need to change as the disparate industries morph into a new, consolidated, and integrated communications entity. These most certainly are interesting times.

In addition to the large, nationwide wired and wireless telecommunications companies discussed to date, there is a rich cornucopia of other players that has affected the telecommunications landscape so far this decade. There are still many small telephone companies that were never part of the Bell System. These so-called independent telephone companies are seeking to survive while not being able to afford the luxury (or necessity) of investing in fiber to the home. Some examples of the mid to high end of the so called independent telephone companies are: CenturyTel, Windstream Communications, Cincinnati Bell, FairPoint Communications, Embarq, and Frontier. Sometimes such independent telephone companies offer marketing plans that incorporate satellite TV and Netflix membership as value added offerings in an attempt to improve their overall value equation, keep their existing customers, and stem the migration to cable competitors. There is also no shortage of companies that are offering WiFi and WiMax Internet access to the masses. In some cases, municipalities have even partnered with them to provide connectivity to their constituents.

Summary

From a telecommunications environment perspective, this decade started out on shaky ground due to irrational exuberance and the Internet bubble and unfounded expectations of great opportunity in the residential telephone local-service marketplace as a result of the Telecom Act of 1996 that opened this frontier up to competition by all. Then it got worse and we observed one of the worst collapses to hit a U.S. industry. From 2000 through 2004, over 650 telecom companies filed for bankruptcy and, at the time, most industry analysts anticipated that it would take at least ten or more years for the U.S. telecommunications industry to recover. Surprisingly, in addition to 2009 marking the twenty-fifth anniversary of the divestiture of the Bell System, it also marks a demonstrably-recovered telecommunications industry that is relatively healthy, growing, and contributing to the overall U.S. economy even though the economy in 2008 and 2009 is in a recessionary state. This period saw the loss of many companies within the telecommunications sector while mergers and consolidations became the norm. We now have only three companies–Qwest, AT&T, and Verizon–that represent the original seven RBOCs plus the original AT&T, MCI, and WorldCom. To a large extent, having large telecommunications carriers seems to make sense as we look ahead. There are great economies of scale available to large horizontally-integrated telephone companies that will morph into overall entertainment and communication companies. For instance, when they negotiate and sign TV content deals with those providers, the larger the base of potential customers, the lower the potential cost to an individual subscriber. Further, when viewing cellular companies, it's clear that having a very large footprint enables a cellular company to offer their services without incurring costs from other cell phone companies, when customers roam as near-ubiquitous service negates the need to partner with others. Passing back the savings to individual customers also makes the cell phone company more competitive and brings more value to the consumer. Bigger is proving to be better in these cases as long as there is not just one powerful provider that can individually control the market and fix prices. Bigger is better, but allowing a provider to get too big and control the market is still unacceptable, as competition yields choice, which usually provides for a fair and market-driven consumer price.

As the telecommunications market recovered, it became clear that there increasingly was a blurring of the heretofore separate markets for television and entertainment and telecommunications. Mobile and wireless add an additional aspect to the mix. Triple and quadruple play market offerings by cable and telephone companies that are morphing into overall integrated entertainment and communication companies is well on its way. U.S. regulatory bodies desperately need to restructure in order to catch up to this newfound and integrated market sector. It's no longer prudent or efficient for one entity to regulate cable for instance and to have another regulate common carriers or wireless. New applications and new, interesting, and fun-to-experience user interfaces are popping up, and the electronic consumer marketplace is beginning to partner with telecommunications providers in an effort to provide even greater convergence and integration. Innovation in technology, policy, security, and market forces abound. Synergies both within and between market sectors are real. The future of the telecommunications sector again looks promising. It is up to regulators and perhaps industry leaders to maintain a reasonable and disciplined approach to this opportunity via sound business practices so that they can avoid the mistakes of the past and move the industry to the next level. They also need to find a way to protect themselves against other companies, markets, and individuals who may not be following these time-tested practices. A common theme and examples of many bust cycles has been: the 2001 telecom market collapse, Internet bubble burst, predatory lending leading to mortgage and financial melt downs in 2008—all resulting in recession. Perhaps an obvious preventive future action would be to prudently limit debt.

Chapter 6 Study Questions

1. Please research and succinctly explain what happened to World-Com Inc.

2. Reflect on the four pillars that support the telecom industry. Identify one event, company, or occurrence not mentioned in this chapter that was of paramount importance during this era.

3. Please research and succinctly explain what happened to Global Crossing during the early part of the twenty-first century.

4. Research the dot-com bubble and provide possible, specific reasons for why it burst.

5. What does the term "network effects" mean?

6. "Traffic-pumping" is one of the less admirable schemes that occurred during this decade. Research and explain 'traffic pumping." What is the current status of this scheme?

7. Ponder the top five dot-com flops. Explain one thing you learned from them.

8. Please explain the term "white knight."

9. Research and explain the difference between an independent telephone company and a CLEC.

10. When it is said that a company is vertically or horizontally integrated, what does this mean? Do companies in the twenty-first century tend to be vertically or horizontally integrated?

11. Some towns and cities have ventured into the provision of wireless Internet. Research this topic and explain the pros and cons of municipalities providing this type of service for their clientele.

12. Interesting and advanced user interfaces, like the iPod, iPhone, mini PCs, tablet PCs, and set top boxes are beginning to offer users better experiences. How does this affect the telecommunications industry?

13. Little is mentioned regarding telephone companies that offer satellite TV by reselling from major providers in an effort to provide triple play services. Consider this approach and briefly discuss if this is a good interim or long-term strategy.

14. Research the issue of major cable providers installing fiber-to-the home facilities. Is this really occurring and if so, where?

15. Research and explain CALEA—the Communications Assistance for Law Enforcement Act. If you hale from another country other than the United States, how does your country address similar issues?

Notes to Chapter 6

95 Spencer E. Ante, "Telecom Back from the Dead," *Business Week*, June 25, 2007.
96 Ibid.
97 Robert Spector, *Get Big Fast*, New York: Harper Collins Publishers, Inc. (2000).

98 Kent German, "The Top 10 Dot-Com Flops," http://www.cnet.com/4520-11136_1-6278387-1.html.

99 Ante, "Telecom Back from the Dead."

100 Lisa Endlich, *Optical Illusions: Lucent and the Crash of Telecom*, New York: Simon & Schuster (2004).

101 Ibid.

102 Ibid.

103 Ante, "Telecom Back from the Dead."

104 Ibid.

105 Wikipedia, "Triple Play (Telecommunications)."

106 Leslie Cauley, "Verizon's Army Toils at Daunting Upgrade," *USA Today*, March 1, 2007.

CHAPTER 7

Crystal-ball the Future!

While all would attest to the fact that currently it's virtually impossible to forecast deep into the future with any degree of accuracy, most also would agree that through the use of such tools as moving average, extrapolation, trend estimation, or possibly regression analysis, econometrics, probability, and simulation, it is sometimes possible to peek into the near-term future with some semblance of accuracy. Even weather forecasters believe that their predictions are often very good for the next day or week, but much less accurate when predicting the weather farther into the future. A futurist is someone who studies the past and present and makes selective suggestions regarding the future. Often futurists discuss the future in terms of the possible, the probable, and sometimes offer "wild card" potentialities that may or may not be currently comprehensible. These potentialities would change the eventual outcomes if they came to fruition. Bright, forward-thinking individuals and companies have come to value futurology or "future studies" so as to better help them somewhat prepare for that time and place that might eventually occur.

The Possible

"Possibility thinkers" care about reality and what they are and what is and they do usually care about accepting the present as the present. However, they also care deeply about innovation and what might be possible. Many consider possibility thinking to be the opposite of dead-end or blocked thinking. It is innovation, creativity, and solutions oriented at its peak. Often, possibility thinking can move things forward by freeing the mover from traps that seem to ensnare such as cynical, pessimistic, and negative thinking. A possibility thinker would seldom, if ever, use

The Thinker by Auguste Rodin[109]

the word "impossible." With this as a preface, please consider the following possibility. Depending on your age, if you were to consider how your ancestors three or more generations ago received their news and information, how would it have been received? It would likely have been through word of mouth. Their children may have received news via newspapers and then via radio. Then came black-and-white television followed by color/stereo TV, and now we may have news, entertainment, and information coming to our homes via HDTV with surround sound, and possibly it may soon be delivered as IPTV over fiber optic cable. First came news via our sound sense and then via newspapers and our visual sense and eventually our sight and sound senses were used together. All this would have occurred within about one hundred years. Now project yourself forward 400 years to the twenty-fifth century. Considering that it took about one hundred years for humankind to obtain news and information via our two senses and become able to enjoy the advances that we enjoy today, who would refute the possibility of obtaining our news and information via more than just sight and sound in the twenty-fifth century? Advances in electronics, genetics, and the biological sciences are immense. Could we experience such communications via all of our senses? Perhaps a twenty-fifth century "All Sensory" device is possible. Just think of it. We could receive news and information via all of our senses: touch, sight, sound, smell, and taste. Perhaps this information could bypass our traditional front-end processors of eyes and ears and go directly to our brains where the information would be processed. If this were possible, then those with sight and hearing challenges that inhibit seeing and hearing would once again be able to experience life on an equal basis as those without disabilities. We would not only have arrived at a richer sensory communications experience, but also a time and place of sensual equality. What a possibility! Perhaps a science fiction writer would take the next step and consider additional senses and/or the ability to transcend time as portrayed in many Sci-Fi stories or even the ability to transcend place via "wormhole" teleportation as portrayed in *Star Trek* and the 2008 movie *Jumper*. It's probably worth asking the question: Where does possibility thinking end and science fiction begin? Likely, George Washington and the rest of the founding fathers of the United States who signed the Declaration of Independence would have considered an eighteenth century writer who wrote about current day television, personal computers, or automobiles to be science-fiction writers. It's interesting that nineteenth century French writer Jules Verne is thought to be the pioneer of the science-fiction genre. His best known works are *Journey to the Center of the Earth* (written in 1864) and *Twenty Thousand Leagues Under the Sea* (written in 1870). "Verne wrote about space, air, and underwater travel before navigable aircraft and practical submarines were invented, and before any means of space travel had been devised."[107] Prior to about 1990, who among us who read the long-running Dick Tracy comic strip and was impressed by the voice-activated video phone that fits around the wrist (introduced in January 13, 1946)[108] would have perceived a real time and place that would allow one to communicate in the same manner as Dick Tracy?

Innovation, creativity, and a vision beyond what is to what might be is an important ingredient toward advancing technology and everything else. This is perhaps the most indisputable portion of "Crystal Ball the Future." Consider how you might break the acceptance thinking of current day reality by envisioning possibilities far beyond mere incremental advances. Innovation, creativity, and the future belong to those who can envision such things! Only you can decide if you have both the desire and wherewithal to become such a "thinker."

The Near Term; the Probable

This near term and probable forecast is more based upon a logical extension and projection of present observations and trends than on futuristic possibility thinking. It's based upon the four major pillars that support the telecommunications environment: technology, policy, security, and market forces, which were first introduced in Chapter 1.

Let us start with the most apparent and well-funded future revolution and change. Some might feel that the most apparent and well-funded change is the wireless phenomenon, but that seems to be more evolutionary. The number one revolutionary change has not really arrived yet for most Americans and is based on the so-called triple- play approach that Verizon, AT&T, and their competing cable companies Comcast, Time Warner Cable, Cox, and others are deploying. The technological breakthroughs of fiber optic technology that allow for virtually limitless bandwidth availability, coupled with the advances of Internet Protocol (IP) and VoIP voice calling, as well as eventually IPTV, are changing the landscape of U.S. market segments from typical telephone based and typical cable-TV based to a converged market segment that will merge and deliver voice, high speed data, and digital TV, as well as other video and more services to United States homes via a small fiber strand. Perhaps we might call this new market segment something other than telephone or cable TV. *Entertainment and Communications Providers* strikes a positive handle while also providing some link to past glories. What name would you suggest for this newly converged market segment that forms from merging the traditional old telephone and cable TV sectors?

Together, today's telephone and cable companies are investing tens to hundreds of billions of dollars by the end of this decade to deliver triple-play services to a relatively modest portion of American homes. These companies are betting that the availability of lots of bandwidth via fiber will be so good and valuable to the home subscribers that their investments will provide a handsome ROI. A positive outcome will yield hundreds of billions of dollars to be available to reinvest and extend the modest coverage of U.S. homes to potentially a ubiquitous U.S. fiber-to-the-home coverage by the end of the next decade or two. What will be the sizzle that causes Americans in their homes to desire this opportunity? First off, the bandwidth will likely be more than 1 giga-bit and perhaps more than 10 gigabit service. This will allow multiple HDTV feeds into these households. Also, there will likely be 100 Mbps or greater Internet service which would provide the speed and benefits of typical business-grade access of today for the home worker, surfer, or gamer. Last, voice communications (and radio) takes such little band-width that it will likely be included in a bundled price that yields the perception of being free for the subscriber. Fiber to the home will put an end to the loss of landlines in favor of all-in-one cell phone plans. It will further incent carriers to provide synergistic services between VoIP calling to one's home and one's mobile communications. Perhaps the first norm will be a simple universal common voice mailbox that is ac-cessible by either the cell phone or landline home phone. So, too, may dual service cell phones become the norm, meaning a cell phone in-strument that can work as a cell phone and also function as an IP phone connected to a WiFi device in the home, thereby reducing the need for air time minutes from the cellular provider. Of course the homeowner would wish to maximize the capabilities of his or her triple-play services by insuring that there is a wireless LAN of sorts available for their laptop, Personal Digital Assistants (PDAs) and now cell phones. Future short-range broadband wireless LANs may be developed so as to be able to easily handle wireless video. One promising option is the USB wireless solution, which claims to offer the speed and security of wired technolo-gy with the ease-of-use of wireless. This would make it possible to move portable TVs around your home as easily as you now move your WiFi-enabled laptop from desk to table to couch. Since there will be plenty of bandwidth available to the home, many applications not yet envi-sioned will be able to easily be deployed. The homeowner at the end of the next decade or two will truly have an enabled smart home and the Entertainment and Communications Providers who participate in this race to provide these services will reap financial benefits for themselves and their shareholders. Those unfortunate "*have nots*," who live in areas that are not within the service areas of these Entertainment and Com-munications Providers, will experience that dread of being "*have nots*." What will eventually become of those telephone and cable companies that don't ride this early twenty-first century wave? Worse, what about the envious homeowners who live in the future wilderness areas that do not have fiber to the home available? Perhaps there will be a "wild card" that will change this trend and potentiality offer nearly unlimited bandwidth without the use of fiber optics, but this often seems improbable. Such a "wild card" might be a wireless solution that provides bandwidth at much greater capacities than today's cellular, WiFi, and WiMax can currently provide. This would negate the need to put the United States on such a high-fiber diet and enable a much less costly deployment of advanced services to the United States. Today, WiMax comes closest to providing the greatest commercially available bandwidth, but it is still much less bandwidth than is available in today's fiber optic networks. Perhaps an Ultra WiMax service will be invented, or perhaps Free Space Optics (FSO) could be harnessed and yield some sort of lower-cost "wild card" that would change the course of triple-play deployment and allow some of today's providers that have less-deep pockets for capital investment to participate in this endeavor. Of course, this could also eliminate the "have not" category. Time will tell.

The wireless phenomenon may not be the greatest and best-funded, revolutionary communications change during the next few years and decades, but it will still be a vast, growing, and very significant future communications change and growth area. Likely, by the end of the next decade, it will turn out to be a wider, more-inclusive, evolutionary telecommunications change than the revolutionary triple-play movement fiber-to-the-home-enabled initiative of Entertainment and Communications Providers.

The biggest driver of change in the U.S. wireless arena is sometimes referred to the "digital dividend." As early as February or possibly as late as June 2009 depending on the channel and its ownership decision, the United States policy makers forced the shut down of all analog television. The movement of all television to digital broadcasting frees up spectrum space previously occupied by UHF channels fifty-two to fifty-nine. While a chunk of this is being turned over to police and fire departments, the auctioning off of the rest provides more total bandwidth to wireless service providers and a financial dividend to the United States through these auctions. As an added benefit, this relatively low frequency–around 700 MHz–is easily able to penetrate buildings well. This fact means that in addition to being good for mobile service, it can work as an alternative to cable or DSL Internet service to homes. Along the way to auctioning off this "digital dividend" bandwidth, the FCC decided to require successful bidders to give customers a much greater freedom in their choice of devices. This telecommunications philosophy, coupled with greater consumer awareness of potential benefits of a so-called "unlocked" cell phone, are together helping to free consumers from the heretofore so-called "walled garden" approach that U.S. cell phone companies had constructed. That is, U.S. mobile providers had restricted the use of their networks to specific cell phone instruments that were locked so as to only be used on their own network. Further, they restricted their network to only allow "approved" applications to be downloaded to them. Can you imagine having to purchase your PC from your broadband provider and only being able to download software approved by it? Surprisingly, this entire concept is not really foreign to the industry. Remember that in the early days, only phones and devices provided by the Bell System were allowed to be connected to the network. Much like the demise of the restriction to only allow Bell phones on the public-switched telephone network, the cellular restriction or "walled garden" approach to cellular will vanish. Likely though, so too will the handsome subsidy that providers allow on cellular phones when one signs a year or two contract in return for a low-cost "locked" phone.

As mentioned in the previous chapter, wireless penetration in the United States is very high (approximately 90%), and the growth rate is tapering off and has dropped to single digits from a rather long run of double-digit growth. Further, the movement to IP-enabled cellular networks and IP-enabled phones will decrease the user's dependence on the cellular network whenever WiFi or other connectivity is available to them. This will somewhat decrease carrier reliance on minutes of usage revenue and pricing schemes. Yet, the health and growth for the U.S. mobile telephone market still looks pretty good. What factors are causing this "good look" phenomenon when penetration is nearing its maximum? Basically three factors are helping to provide a positive short-term forecast: the enabling ability of increased bandwidth via the "digital dividend," the movement to IP and everything digital, and advances within consumer electronics. In addition to these basic three near-term factors, there will be the "forklift upgrade" movement from Third Generation Mobile Communications (3-G) to Fourth Generation Mobile Communications (4-G) some time during the later part of second decade of the twenty-first century. Consumer willingness to pay for the 4-G features and services will be pivotal to the roll-out of 4-G in the United States.

Greater bandwidth and the "digital dividend" have already been discussed. In effect, they provide a larger building lot or a bigger, richer, and more fertile sandbox within which to grow. The movement to IP and everything digital, as well as advances in consumer electronics, are complementary to each other and will help to provide the new and innovative applications and structures that will fill up the larger lot and sandbox that greater bandwidth enables. Everything IP and digital helps to provide fertile ground for interesting and useful new applications to develop and the microelectronics within consumer electronics can further enhance these applications and make the user interface small enough to be used for mobile and interesting enough to provide a certain "wow" factor that enhances the user experience. Perhaps the introduction of the iPhone during the summer of 2007, followed by the introduction of iPhone 3G the following year, is a perfect example of this phenomenon. If you aren't familiar with this device, please become familiar with the iPhone 3G by answering end-of-chapter study question 7.

Just what are some of the new, interesting, and useful telecommunications-delivered applications that are or might enhance our lives? Obviously, there are way too many to discuss them all here, but two of the more important nonentertainment applications are Global Positioning System (GPS) related and sometimes referred to as location-based services and also directory-related services.

A location-based service marries GPS or exact location positioning with Geographic Information System (GIS) data including addresses and various other terrain information. Location-based services are fairly well developed and are beginning to penetrate users such that they are one of the more popular advanced mobile applications. Consumers can expect to see improvements in several categories of location-based services including: navigation, search, real-time information, and social networking.[110] Navigation via turn-by-turn directions is the most popular of this category of services, and the industry is working to further refine it. Google's "My Location" on Google Maps can detect your location based on cell towers, and Google maps also launched a feature that provides public transportation options to your destination. Perhaps the cost of gasoline may some day make this an even more sought-after application. Navteq is working on a real-time map for walkers that would feature more granular details and information, perhaps with visual cues that help one distinguish locales. A Google Android application called "BreadCrumbz" uses satellite images and other street photos to provide directions. Moving beyond navigation services to the search function, location-based services can provide individuals with the nearest pizza, restaurant, hotel, or hospital. Another Google Android application, "Wikitude," can provide Wikipedia descriptions of points of interest near the user so that one doesn't miss the most significant points of interest while in that location. Now if Google could expand that application to include not only places of interest, but also people and other things, we would all miss less of the important things in life. Traffic alerts and updated weather conditions are very popular real-time applications, as are flight schedules. There also seems to be development in merging location-based services with popular social networking tools. Loopt's "friend finder" can allow you to track down friends and family members as they move about. The entire concept of linking location-based technology with other applications, like social networking, appears to offer bountiful ground for the next advanced-life-enhancing- communications-provided experience.

Directory-related services are the other interesting and potentially useful class of applications that could well enhance the lives of U.S. individuals and potentially the rest of the world, if only more research into proper methods and policy guidelines that safeguard the individual's privacy could be agreed upon and developed. After all, other than for e-mail, search is often the most used and useful Internet tool. There are probably hundreds, if not thousands, of search algorithms. New, faster, and more efficient methods and algorithms are constantly under development. Why not use this modern, efficient, and most useful Internet tool to find telephone-related information, including residential telephone numbers — all residential telephone numbers? Clearly, the potential is apparent, but due to lack of guidelines or possibly consumer desires, it is a NOT-ready-for-prime-time application for residence numbers when cellular is included. Not done correctly, there is potential for this set of applications to become a nuisance, if not outright harmful. Directory listings information is not really a new application as it relates to telephone numbers, but rather an old application with the modern twist of being online and potentially being able to address all residential telephone numbers including cell phones. Unfortunately, valid concerns for privacy and also the methodology of charging a recipient cell phone for calls placed to it have substantially stalled this important application as it relates to residential telephone numbers, so this application is not far along in development and penetration. Nevertheless, it has great potential if done well and with the proper guidance from regulators in order to protect one's privacy.

Consider the local telephone directory. Usually one has either one directory divided into "white pages" for residential telephone listings and "yellow pages" for businesses, or there is a separate directory for each category. Just about every business experiences great advantages by being listed in the "yellow pages," and this is why the businesses actually pay the directory provider to advertise their number. Directory advertising is an efficient and proven positive method of advertisement. Online fixed and/or mobile directories for business are a natural positive applications category today and will flourish in the future. Directory assistance is currently about an $8 billion-a-year business. While this is not trivial, what could it grow to by moving it online and providing both wired and wireless access? Perhaps Google and Microsoft and others have the answer, as they're among several companies that are entering this market. Dial 1-800-GOOG411 and Google will connect the call and text the number to the user's cell phone — all free to the business directory searcher.[111] Residential directories are another story. Most of us recognize the value of an alphabetical listing of land-based residential telephone numbers. Individual residences typically don't pay the directory provider for listing their phone number, but rather must pay to have their number *not* listed; thereby becoming what is called a "nonpub" number. So-called "nonpub" telephone numbers are not only not published in the paper telephone directory today, but when one calls directory assistance to find someone's telephone number that is "nonpub," the directory assistance operator is not allowed to release the number since the "nonpub" subscriber has paid for this privacy. Nonpub numbers are also not available in online telephone directories. Currently, all U.S. cellular telephone numbers are not published or listed, and there is no charge to the subscriber for this condition. To further complicate this issue, cellular users currently pay for air time when they are called in the United States. In many, if not most foreign countries, it is the calling party who pays for a cellular call and not the called party. If individuals and telemarketers were able to look up cellular telephone numbers in a cellular directory and call the numbers, U.S. cellular users would not only be interrupted by such a call but they also would be charged for it, thereby adding financial injury to inconvenience. To prevent harming an individual's privacy and to prevent undesired calls from accessing cellular residential telephone numbers and causing charges, new approaches to this challenge need to be developed. Cultivating the good while preventing the bad through the use of modern-day tools would appear to be doable and potentially worth the effort. Possible considerations might be the use of multiple IP addresses for every telephone—a public and available address coupled with a private and nonpublic address. The use of the concept "presence" to quickly provide first search results based on the presence of the device making the query could be used to improve efficiencies of the search, while programming techniques could likely be deployed that would implement whatever privacy guidelines are finally adopted.

There are really two bottom lines to this directory application. For business, both the old and the many new approaches to deliver this application are and will continue to be deployed. The old (paper) approaches should be preserved so as to facilitate the use by those without online access through choice or circumstances, while new online wired and wireless approaches will flourish for the majority of technically proficient users with capable online access. The directory application for residential telephone numbers, however, is problematic. Residential directories have value, but if no action is taken, then current residential directories will diminish in value as many users forsake landlines in favor of only having a mobile (not published) telephone number that results in an increasingly more incomplete residential directory. As we look to the future, industry and/or regulatory leadership is required in order to research and possibly develop the safeguards that would be necessary to provide a comprehensive useful residential telephone directory while preserving individual privacy for both wired and mobile telephone users. Are the benefits of such a directory worth the leadership, research, and implementation investment? The current lack of traction regarding a comprehensive residential telephone may be telling.

Closing out the infusion of extraordinary wireless initiatives is the movement to Fourth Generation Mobile Communications some time closer to 2020 than to 2010. It will provide that next impetus for advancing mobile telecommunications both in the United States and globally. In fact, it will advance all telecommunications, since it will offer the enabling abilities available though greater bandwidth and also be all IP based, which together will provide for excellent integration capabilities with land-based communications and entertainment systems. G-4 standards should finally yield global compatibility so that a cell phone will operate anywhere in the world and advanced multi-media, including Web applications and video, will be the norm. Within the U.S., customer perception of the value that can be derived from 4-G will be pivotal to its deployment timeline. Providers must be able to project customer acceptance beyond very good 3-G mobile systems and applications so as to obtain a good ROI. There are parts of the world that have not deployed 3-G mobile in favor of moving from 2-G directly to 4-G. Likely, these areas will be the earliest to implement 4-G.

In addition to the revolutionary thrust of fiber-enabled, triple-play, land-based residential communications systems and the tremendous evolution of wireless communications, there will be a rich and extraordinary amount of other advances, innovations, and changes. The following are a few of them:

- Businesses will continue to migrate their voice-calling networks to VoIP and replace their old circuit-switched TDM PBX with VoIP soft switches, if they haven't already done so. Residences will continue their migration to VoIP as well. Initially, these

VoIP migrations are driven by the better economics of VoIP over circuit-switched telephony, but, interestingly, advanced capabilities of all digital IP communications will allow for many new and advanced features and options as well as enable the complete and smooth integration of disparate networks and communications mediums, thereby enhancing connectivity, communications, and our lives. Cellular carriers also plan to convert their networks to all IP.

- Somewhat analogous to the many benefits offered by the fiber to the home residential initiative will be the "Carrier Grade Ethernet" fiber access benefits for medium and small businesses that heretofore couldn't partake of fiber due to their limited size. Only large corporations, universities, and their providers were easily able to justify the cost of fiber and its benefits in the past. Just as providers like AT&T and Verizon envision positive ROI for fiber to the home, they and others envision a positive ROI for "Carrier Grade Ethernet" rings that will deliver the fiber benefits to medium and small business markets.

- There will continue to be a large influx of nontraditional telecommunications companies entering or in some way affecting telecommunications, or should we say Entertainment and Communications environment? Google and Microsoft are two examples of large and influential companies that will continue the trend of penetration into this market by nontraditional companies. There are and will continue to be many, many more such entrants. One very specific and even somewhat traditional applications example of Microsoft's involvement is their Microsoft Unified Communications (MSUC) product, which directly competes with traditional telephone and Internet-type company products. The intent of this product is to provide a platform that not only supports the provision of multiple services, but also integrates with existing networks and takes advantage of existing investments, such as MS Exchange Server. It includes the following:[112]

- MS Office Communications Server — Software that delivers VoIP, video, IM conferencing, and presence within applications users already know such as MS Office.
- MS Office Communicator — Client software for the phone, IM and video communications that works across the PC, mobile phone, and Web browser.
- MS Office Live Meeting — An advanced conferencing service that enables meetings, sharing of documents and video, and is able to record the meeting.
- MS Round Table — A conferencing phone with a 360-degree camera.
- MS Exchange Server — an industry leading e-mail, voice mail, calendaring, and unified messaging platform.

- There will also be many subtle social changes that will often be imperceptible until one examines them further.
  - One such change is the change from place-to-place calling in favor of person-to-person calling for nonbusiness communications. One typically no longer places a call from one's residence location to that of another, but, rather, makes a call from his or her cell phone while located just about anywhere and calls an individual on his or her cell phone. While voice mail often gets in the way of that direct connection, the communications link is really person-to-person unless the called party is busy or is otherwise not accepting calls.
  - Another subtle social change is the movement of personal communications intended to share feelings or establish a mood, relationship, or social link rather than to merely communicate information and ideas. The word that best describes this is **phatic** communications. A large percentage of nonbusiness cell phone use involves phatic communications.

- "The first-ever 911 call was placed in Haleyville, Alabama in 1968."[113] This was the beginning of the wireline 911 introduction in the United States. This nationwide universal emergency assistance number and system has long been recognized as highly important. "This importance was acknowledged with the passage of the Wireless Communications and Public Safety Act of 1999. Subsequent developments, e.g., the tragic events of September 11, 2001 and growing dependence on wireless networks, serve to further emphasize the importance of E911 in general, and wireless E911 in particular, to the safety of life and property and homeland security. The automatic provision of location information with wireline and wireless 911 calls is critical to emergency services."[114] Unfortunately, "while generally reliable, it is seriously antiquated."[115] It turns out that "the existing wireline E911 infrastructure is built upon not only an outdated technology, but one that was originally designed for an entirely different purpose. It is an analog technology in an overwhelmingly digital world. Yet, it is a critical building block in the implementation of wireless E911 and IP telephony."[116] Due to these facts, the FCC has taken a national policy leadership role in coordinating the many diverse stakeholder entities in an effort to properly address all phases of this issue, including cooperation among stakeholders as well as proper funding for the large additional quantities of calls now directed to Public Safety Answering Points (PSAPs, a.k.a. E911 centers). During the summer of 2008, PSAPs in Rochester, NY; Bozeman, Mont.; King County, Wash.; St. Paul; and Fort Wayne, Ind. began testing the Department of Transportation's Next Generation 911 system with the goal of replacing the four-decades-old technology that now governs how the nation's 6,000-plus 911 call centers operate. Capabilities will include the acceptance of input via: voice, text messaging, VoIP, mobile, and navigation services, such as On Star. "It could take eight to ten years for all of the nation's 911 centers to upgrade to the new technology. An analysis by Booz Allen Hamilton estimated the cost of the Next Generation system could be less than the current one over the next twenty years. It estimated the system to cost $57 billion to $64 billion in that span compared to $55 billion to $79 billion for the current system."[117]

- What do we call an advance that is not forecasted, but would yield a better world if it became a reality? Perhaps it's a missed opportunity, oversight, or even a missing priority. Perhaps it's an unfilled need. In any case, one such item is the need for greater leadership and industry responsibility relative to setting up an important, unbiased, and comprehensive, public and open system and process to protect the public relative to health and safety issues in the Entertainment and Communications environment. Few would be able to successfully argue that if there had been a better system to address the health and safety issues of tobacco, smoking wouldn't have wounded and killed so many Americans. It's unfortunate that some growing world economies have not learned from the ill effects of smoking in the United States, as tobacco smoking is still prevalent in many parts of the world even though it's a known carcinogen and will harm individuals who smoke and/or are exposed to smoke. The world also has finally awakened to the previously unanticipated ill effects of not properly caring for the environment. It would be difficult to argue that "global warming" is not a concern. Hopefully, it is not too late for mankind to somewhat reverse the cumulative effects of global warming and other environmental determents. These and other potentially harmful activities could have been addressed much sooner had there been greater leadership and industry responsibility relative to setting up an important, unbiased, and comprehensive, public and open system and process relative to health and safety issues, including standards. The entertainment and communications environment needs such a system. Most individuals believe the studies that indicate that Radio Frequency (RF) radiation's affects on humans are negligible as they've shown that there is no proven harm. Somehow, it seems that a safer statement would be that studies have proven that under X, Y, and Z conditions, RF radiation has been proven to be safe to humans or at least as safe as other typically-used products. This is a very complex issue and many could argue that it is only possible to indicate that a products use yields no proven harm. Perhaps future technologies will pose similar dilemmas relative to health and safety. At the global level, the World Health Organization is involved. Within the United States the FCC, Food and Drug Administration (FDA), and others are involved. Are they doing enough? Possibility thinking leads one to believe that it's possible to do better. Why not prioritize the creation of a system and process that will address health and safety issues so as to prove them safe? The entertainment and communications industry would be better and visibly able to demonstrate it by creating a process and system to adequately address health and safety issues that could affect the industry and users of its products and services. At this juncture, this outcome is not yet forecasted.

- As we look to the future, there are many entertainment and communications products and services that have, or may in the future, "bite the dust," because they have no future.
  - Certainly the rotary dial telephone, old generation cell phones and analog cell phones, 8-track tape, cassette tape, VCR, and soon the DVD may become more prevalent in museums than in actual use. It's currently possible for the oldest telephones, including Bell's first devices, to function on today's Public Switched Telephone Network (PSTN) without adapters. That will change over the next few years as the United States moves to IP telephony and eventually the PSTN as we know it ceases to exist and then only IP- enabled phones will function.
  - At the end of 2007, AT&T announced that, after 129 years, it was discontinuing its pay phone business. "The first pay phone was installed in 1878 and had an attendant who took callers' money. William Gray set up the first coin-operated phone in 1889 at a bank in Hartford, Connecticut. At the peak in 1998, there were 2.6 million pay phones in the United States. That number has fallen sharply to 1 million including the 65,000 phones distributed over thirteen states that pioneer AT&T had left at the end of 2007."[118]

- In the early days of the telephone — in fact, for almost the first one hundred years — it was only possible to purchase one's telephone from the provider, usually a Bell System Company. Through the end of the twentieth century and into the twenty-first century, it was necessary to purchase one's cell phone from the cellular service provider. This will no longer be the case as unlocked phones, coupled with a more knowledgeable and savvy user, are forcing this change.

- If nearly every college student has a cell phone, what benefit is there to colleges and universities investing in and maintaining telephones in dorm rooms? At one time revenue derived from renting such a device and providing local and long-distance service to students was an asset for schools and students received a desired service. Now it is a cost liability for schools and this is passed along to students even though most students neither need nor desire a wired phone in their rooms. Perhaps a phone in each hallway would suffice?

- There are many more products, services, policies, regulations, and laws that will change.

Summary

Forecasting the future with some degree of accuracy is challenging and problematic at best. Considering that almost anything may be possible, "possibility thinkers" are developing creative innovations and moving beyond mere logical extensions of what is to what might be. Surely this is important! So, too, is the less-demanding forecasting based upon a knowledgeable and logical extension of current observations and trends. This yields a believable and fairly accurate near-term forecast of the future.

The results are that the near term will see much focus on the consumer. Fiber to-the-home will allow advanced triple-play entertainment and communications initiatives that promise excellent ROI for the carriers and a cornucopia of services for the home and home office consumer. These services include multiple HDTV and/or IPTV feeds, very- high-speed Internet service, and, of course, voice calling via VoIP.

Homeowners will have wireless LANs that can enable connectivity to most devices through advanced wireless connections, which will include wireless IP connections to cell phones, thereby decreasing their reliance on and use of cellular telephone network minutes whenever near such a wireless access point. Homeowners and small office/home office (SO/HO) users who do not live in fiber-enabled neighborhoods will become the "have nots" of the future unless there is some sort of "wild card" technological breakthrough that will provide for wireless bandwidth availability on par with fiber. Without such a breakthrough, the advanced products, services, and benefits that the "haves" enjoy will not be theirs. It is said that the most valued and important attribute of real estate is location, location, location. This scenario adds a new twist to the location of a home.

While wireless penetration in the United States is approaching its peak, wireless will also greatly grow and expand in terms of usage, applications, and value. There are multiple drivers for this phenomenon. First there is the so-called "digital dividend," which helps to expand available wireless bandwidth in somewhat of an analogous way that fiber expands available bandwidth for wired communications. Everything digital and IP, and advances in consumer electronics, will also fuel this engine in the short haul, while the movement to Fourth Generation Mobile (G-4) will provide added adrenalin later on. Always remember that while the technology and delivery methods are important enablers, it's the capabilities for the user through advanced applications that are really the key to growth and development. Applications such as location-based services, directory and lower-cost voice calling through IP are examples of key applications. So, too, are business applications that integrate and greatly increase knowledge worker productivity, such as Microsoft Unified Communications (MSUC). The modernization and implementation of a nationwide universal emergency assistance system is a top priority, application and initiative too. Fortunately, there are many, many new and developing entertainment and communications applications that will add value and enhance lives. As some of these improvements occur and become proven by being assimilated into everyday life, many of the old applications and systems will be phased out due to obsolesce. Perhaps the time is ripe to add relics to telecommunications and cable museums?

If one were to step back and reflect on this near-term forecast for the entertainment and communications industry and its users, perhaps the most important lesson is that it is a very uplifting and excellent forecast. Truly, we live in interesting times!

Chapter 7 Study Questions

1. Please do a little focused research regarding Aristotle's ethical philosophy of knowing the good. Unlike the many great philosophers before him, including his teacher Plato, he believed that it was not enough to merely "know the good." Once one knows the good, one must "do the good." Why is this important? How does this fit in with the study of telecommunications?

2. Why, in your opinion, does the author feel that the move to triple-play services and fiber to the home is a greater change than the wireless revolution?

3. Wireless USB is mentioned in this chapter. Compare and contrast the following: Wireless USB, Bluetooth, WiFi, and WiMax.

4. What is a "locked" cell phone?

5. Explain the "walled garden" approach that U.S. cellular companies have used in the past.

6. Research and define the main difference between the so-called "digital dividend" in Europe vs. the United States.

7. Research Apple's iPhone G3 and explain why it is a perfect example of the marriage of advances in applications, IP, and digital electronics. What might additional bandwidth for its network provide?

8. Research and consider any negative aspects of location-related services and explain why they are potentially negative. How might concerns for these be addressed?

9. Research the various generations of Mobile communications and briefly explain them.

10. What awe-inspiring, nonvoice applications will likely make 4-G Mobile come to market sooner rather than later?

11. IMS-IP Multimedia Subsystem was specifically not mentioned in this chapter. What is the vision of IMS and, in your opinion, will it become popular and benefit evolving mobile networks?

12. During the very early years of telecommunications, universal service proved to be a positive policy and force for both users and AT&T. At the time, it helped make telecommunications in the United States to be the best in the world. Today, voice communications within the United States is universal. Consider whether or not universal service should be extended to include Internet access. Please explain the probable positive and negative aspects of such a policy. What about universal electric service?

13. The use of satellite is an interesting approach to providing entertainment services. Does satellite provide an opportunity to eliminate the concept of "have nots" that might eventually occur when half of the United States enjoys fiber-to-the-home and the services and benefits that it will provide?

14. What additional products, services, or policies are vanishing or will soon vanish?

15. What additional future items would you include in this chapter?

16. Please provide any other suggestions that you have for considering the future of telecommunications in the United States.

Notes to Chapter 7

107 Wikipedia, "Jules Verne," Wikipedia, http://en.wikipedia.org/wiki/Jules_Verne.

108 Alan Chodos, "Dick Tracy Watch," http://www.aps.org/publications/apsnews/199906/dicktracy.cfm.

109 Wikipedia, "The Thinker," Wikipedia, http://en.wikipedia.org/wiki/The_Thinker.

110 Roger Yu, "GPS Becomes a Vital Tool for Frequent Travelers," *USA Today*, July 8, 2008.

111 Olba Kharif, "Free Information Please," *Business Week*, April 23, 2007.

112 "Microsoft Unveils Unified Communications Product Road Map and Partner Ecosystem," http://www.microsoft.com/presspass/press/2006/jun06/06-25UC-GRoadmapPR.mspx.

113 Dale Hatfield, "A Report on Technical and Operational Issues Impacting the Provision of Wireless Enhanced 911 Services," 2002.

114 Ibid.

115 Ibid.

116 Ibid.

117 Matthew Daneman, "Five Cities Test High-Tech 911 System," *USA Today*, July 9, 2008.

118 Crayton Harrison, "AT&T Disconnecting Pay Phone Business," *Democrat and Chronicle*, December 4, 2007.

CHAPTER 8

The End of the Beginning: A New Beginning!

History tells us that communicating at a distance has been both a desire and a need of mankind since the beginning of human societies. So-called modern telecommunications began with the invention of the Telegraph in 1838 or the invention and highly successful patenting of Bell's telephone in 1876, depending on one's perspective. Regardless of perspective, though, the decision by the U.S. government to not purchase the telegraph patents from Samuel Morse yielded an environment in the United States in which telecommunications developed in the private sector rather than the governmental or public sector of the economy. This was a global anomaly and one that enabled a unique telecommunications model to evolve and to remain unique for over one hundred years, while the rest of global telecommunications markets were also evolving. This unique attribute makes the U.S. model well worth study. Today's and likely the future's global telephone markets will grow and evolve in the private sector and perhaps someday be more regulated by the invisible hand of free trade market forces than by traditional regulatory bodies. In this regard, the United States and global telecommunications markets are converging. Telecommunications or entertainment and communications environments both in the United States and globally, are highly affected by technology, policy, market forces, and security.

When Alexander Graham Bell invented the telephone in 1876 to address a communications desire and need, he also coincidently launched the telephone industry. This industry went from the early years of Bell's company having little or no competition due to patent rights on the telephone to intense competition and chaos soon after those patents expired. There was then a period of over fifty years of a regulated monopoly model in which technological and market leadership, as well as much of basic telephone service, was provided by one large corporation called AT&T and which was often referred to as the Bell System. After some early years of noncooperation that was corrected by the U.S. government, the Bell System eventually interconnected with the large number of small, independent, non-Bell telephone companies and they all worked together to provide excellent nationwide telephone service and arguably the finest telecommunications service in the world. Considering telecommunications around the world, some would argue that the U.S. telecommunications excellence provided U.S. business and industry with an added advantage relative to their global competitors. Advances by others have since eliminated this U.S. competitive edge.

Interestingly, approximately midway through the twentieth century and with little regard to having the best telephone service in the world, the U.S. government and many individuals and corporations began to challenge the concept of a single-source provider, as well as the concepts of protected markets and the resulting monopoly this provided the Bell System. From a global perspective, this challenge was unique to the U.S. telephone industry. At the time, the telephone industry was often characterized as a natural monopoly. Public opinion and public policy moved from that point in time when the provision of telephone service via a regulated monopoly was thought to be good to a point in time when the competitive model became king. The breakup of the Bell System, which is commonly referred to as "divestiture," was a pivotal action aimed at moving the U.S. telecommunications industry toward a more competitive model. This consent decree was initiated by the U.S. Department of Justice (DOJ) and agreed to by AT&T and was effective on January 1, 1984. This action tore the Bell System apart by creating seven RBOCs directed to provide local and intraLATA telephone service and AT&T, which provided interLATA or long-distance service. The Bell System was separated in this manner in order to eliminate the "power and incentive" to discriminate, thereby ushering in a more competitive model for long-distance service and fostering the growth of the IXC marketplace in which the MCI Corporation became a significant competitor to AT&T. Divestiture was highly successful in bringing competition to the industry and in lowering long-distance rates to all consumers.

While the long-distance telephone market became highly competitive after divestiture, the local service arena remained a closed marketplace. Local telephone exchange carriers pretty much had a monopoly over it through the 1980s and into the 1990s. After years of debate, U.S. lawmakers finally reached consensus by crafting the Telecommunications Act of 1996. This was a sweeping revision to the Communications Act of 1934 and was directed to provide competition to all remaining telecommunications markets, as well as move toward deregulating the industry and accelerating private sector deployment of advanced telecommunications and information technologies for all Americans. The most successful part of this Act was that it succeeded in bringing forth the notion that all telecommunications markets should be open to competition by all. However, the economics of providing local telephone exchange service as envisioned in the 1990s was much different than that of implementing long-distance in the 1980's so the Act's resulting effects on competition were not so successful. This is especially true relative to the tremendous positive effects that divestiture had on the long-distance marketplace the decade before. However, as it turned out, advances in telecommunications technologies, such as the Internet, IP, fiber optics, VoIP, and cellular communications helped to accomplish local telephone exchange competition in ways that were not even envisioned by the Act. What the policy of the Telecommunications Act of 1996 didn't accomplish, due to the economics of the marketplace, was eventually accomplished, and more, by technological advances.

The year 2000 ushered in a new decade and a new millennium. Unfortunately, this started out on shaky ground due to irrational exuberance, and the Internet bubble and unfounded expectations of great opportunity in the residential telephone local-service marketplace as a result of the Telecom Act of 1996. Then it got worse and we observed one of the worst collapses to hit a U.S. industry since the Great Depression. (The worst to date being the mortgage, banking, and financial collapse of 2008) Fortunately, the milestone that we're passing marks a turnaround and recovery from the depths of 2001 that is far greater and sooner than anticipated although it is muted a bit by the 2008-2009 recession. As the recovery unfolded, it became clear that there was an increase in the blurring of the heretofore separate markets for television and entertainment and telecommunications. Triple-play market offerings over fiber optic cable are morphing these former distinct markets and industries into one which, until we come up with a better label, we'll call the Entertainment and Communications Industry. At the same time, cellular wireless is nearing its peak saturation levels, but innovative applications and interesting, new, fun-to-experience user devices are being launched, thereby offering expanded growth potential. As so many advances occur and are woven into everyday use, old applications, devices, systems, and notions will be phased out due to obsolescence. While the events of the past have been significant and important, the future will be much different than the past and the best is yet to come. The industry has evolved from a single application (voice) in the earliest days to multiple applications, delivery methods, and modes today. Who knows what tomorrow may bring? The industry is richer and much more interesting and offers greater opportunity and fulfillment than ever before. Perhaps the time is ripe to close out or end the old concepts surrounding telecommunications and consider this future Entertainment and Communications age to be A New Beginning.

Let us close by revisiting a question that was put forward at the beginning of this book. "Why are the history and evolution of telecommunications and the forces that have affected the telecommunications environment important?" Professor, administrator, and trained historian Dr. Laurence Winnie is currently at Harvard University. He believes the concept and words that best address this question are found in a quotation of German philosopher Odo Marquard who said, "Zukunft braucht Herkunft[119];" or in English, "The future needs provenance." Winnie goes on to say ,"Just as an object or a work of art needs *provenance* — a record of in whose hands it has passed, or where it was from — so does the future. Even the future needs its sources — its origins — in order to decipher the mystery of its content. We cannot know what the future is, even as we experience it as the present, without the past. The future comes out of the past. Thus, thinking backward, features of the future may be, at least, expected from the past and around us in our present, even if they cannot be seen in precise detail." Of course, you know this!

Chapter 8 Study Questions

1. Are you from a country other than the United States? If so, using this brief chapter as a guide, write a four page summary of your country's telecommunications environment.

2. There were few specifics mentioned regarding the fact that security is a major influence to the telecommunications environment. Please comment on the security leg of the four-legged stool and suggest what you would add to this text regarding security and where you might make such additions.

3. Considering that divestiture broke up the Bell System in the manner that it did in order to separate the "power and incentive" to discriminate, why then were mergers and takeovers allowed so as to yield the new AT&T and Verizon?

4. Genealogy may be part of the provenance of individuals. Please comment on this concept.

5. How might the new Entertainment and Communications sector and offerings change the way you live?

Notes to Chapter 8

119 Odo Marquard, *Zukunft Braucht Herkunft* [The Future Needs Provenance], Stuttgart, 2003.

Case Studies

This book has endeavored to document, demonstrate, and validate how technology, policy, market forces, and security affect the telecommunications industry and environment. Attempts have been made to incorporate specific examples and testimonials along the way. However, the telecommunications industry and environment are rather broad in nature. It should be pointed out that broad and complex entities like these may affect and be affected in many different ways. So, too, may specific corporations and individuals be impacted in distinctive, special, and often dissimilar ways. This appendix provides a granular view of selected contributors to the telecommunications environment by focusing on case studies and issues and some specific corporations. How they may have been affected and impacted and perhaps affect the telecommunications industry and environment is discussed. Not all the ramifications and answers to all questions will specifically be revealed in each case, but through individual student research and classroom discussion, it's anticipated that students may learn much through this appendix and approach.

PAETEC

"Though you've probably never heard of PAETEC Communications, Inc., by most measures we're the most successful telecom of the past decade. As a full-service telecommunications company based just outside Rochester, New York, we come out on top in almost any comparison against other little companies you don't hear much about, and the big guys you may wish you'd heard a little less about ..."[120] These comments were taken from the introduction of *It Isn't Just Business, It's Personal* written by PAETEC cofounder Arunas A. Chesonis in 2006. This was before PAETEC's mergers and acquisitions of US LEC, Allworx, and McLeodUSA, which took place in 2007 and 2008, and before it became a public traded company. Today, PAETEC has emerged as the country's largest Competitive Local Exchange Company (CLEC) and top competitive provider of local telephone service. At the turn of the millennium, there were about 330 such competitors like PAETEC, but by 2009 there are only about twenty-five left. Why did PAETEC succeed when so many other CLECs failed in their attempts to seize the opportunity that presented itself as a result of the Telecom Act of 1996? The purpose of this case study is to address this question by learning about PAETEC Communications, Inc., its path to success and why it is successful, and where it may be headed.

A January 2009 snapshot of PAETEC's Web site provides a succinct view of its founding principles, business, and, corporate mission and values:

> **Founding Principles:** PAETEC was founded in 1998 when CEO Arunas A. Chesonis and a core team of executives recognized that existing telecommunications providers were failing

to respect the most important factor governing the long-term success of any business — first-rate customer service.

Who We Are: Today, PAETEC delivers personalized communications solutions and unmatched service to business-class customers in more than 80% of the nation's top one hundred metropolitan areas. We are the premier alternative to the ILECs, based on our nationwide footprint, breadth of products, and quality of service.

PAETEC's data and voice products — and our unique value-added offerings — help customers achieve cost-effective solutions.

Corporate Mission: PAETEC's corporate mission is to be the most customer- and employee-oriented communications provider.

Corporate Values: Our dynamic growth has been achieved by adhering to basic values that will continue to define PAETEC in the future: The essence of the PAETEC Experience can be summarized in the following four Corporate Values:
- Caring Culture
- Open Communication
- Unmatched Service
- Personalized Solutions

Every aspect of our company is aligned with at least one of these four values, whether it is how we run our business, satisfy our customers, or treat our people. There are many reasons why customers initially select PAETEC; however, the relationship established is what keeps them with us.[121]

Additional PAETEC facts are:

PAETEC is headquartered in Perinton, NY, which is a suburb of Rochester, NY. Annual revenues after purchases and mergers are approximately $1.6 billion with Earnings Before Interest, Taxes, Depreciation, and Amortization (EBITDA) estimated to be approximately $230 million per year for the combined entity of PAETEC, US LEC, Allworx, and McLeodUSA. Total phone lines, which are all business lines, are approximately 3.345 million. PAETEC's stock hovered in the $1.50 range as 2009 was ushered in, which is considerably below its high $12+ stock price per share when the purchase of McLeod USA was announced in September 2007. Some might argue that this is a sign of potential weakness. It is true that it's a sign of weakness, but not so much for PAETEC. Rather, it's a sign of the weakness of the entire U.S. economy during 2008, which turned out to be the worst year on Wall Street since 1931. Lowlights of 2008[122] included:

- $6.9 trillion in stock market wealth being erased — roughly 38% of the total value of U.S. stocks at the start of 2008.
- The average price of a share listed on the New York Stock Exchange fell 45%.
- The Standard & Poor's 500 index — the indicator most watched by market pros — fell 38.5%.
- The Dow Jones industrial average dropped 33.8%.

Broad markets dropped and, of course, individual corporate stocks compose these markets and their stock prices decreased too. Even Apple — the electronics juggernaut with the iPod and iPhone — saw their stock plunge 57% in 2008.[123] Suffice it to say that PAETEC, like nearly all other U.S. companies, was not immune to the devastating economic tide of 2008. However, in difficult economic times, it is likely that a company such as PAETEC, that offers superior service at value prices, may actually fare better than some of the giants of the telecom industry, and so the stock price resulting from the 2008 economic difficulties is not really at issue. Yet, we should never forget that positive economic times are always best and more pleasant, because the overall economic pie expands, and it's far better to get one's market share from this larger pie. Perhaps the U.S. economic stimulus package of 2009 and 2010 will expedite a return to such times.

Let us learn about the path that brought PAETEC to its current status beginning with some background information on its founder Mr. Arunas Chesonis. He began his career at Rochester Telephone Corp in the 1980s where he gained a sound knowledge of both telephone technology and the industry. Rochester Telephone is now named Frontier and is owned by Frontier Communications, formerly Citizens Communications Corporation. After leaving Rochester Telephone, "he went on to serve as President of ACC Corp., the parent company for all ACC-owned operations in the United States, Canada, Germany, and the United Kingdom, from February 1994 until April 1998, and was elected to its Board of Directors in October 1994."[124] He founded PAETEC in 1998. Note: Teleport actually purchased ACC in 1997, and then AT&T purchased Teleport/ACC in 1998 and made the then chairman and chief executive of Teleport, Robert Annunziata, an AT&T Executive VP and head of their new local services unit. This unit was then a promising new AT&T initiative. Chesonis holds a BS in civil engineering from Massachusetts Institute of Technology (MIT), an MBA from the University of Rochester's (U of R) Simon Graduate School of Business, and an honorary doctorate of law from the University of Rochester. He is chairman of the Earth System Initiative at MIT and serves as trustee at the Harley School, RIT, and U of R. Let us add this note regarding the derivation of the PAETEC name. It actually comes from within Chesonis' family members — his wife, Pamela, and children Adam, Erik, Tessa, and Emma Chesonis. Until 2004, the company was PaeTec with the T capitalized.[125]

The birth of PAETEC is actually the result of a unique confluence of market forces and telecommunications policy changes coupled with what, in retrospect, seems like perfectly orchestrated timing. Likely, if one were purposely attempting to create a perfect incubator for such a creation, it would never happen. When AT&T purchased Teleport/ACC, it created an opportunity of "*buy out*" for some senior management of ACC. In addition to this opportunity, it provided a disincentive for any true entrepreneurial type of individual to stay on and manage under the AT&T umbrella. Thus, Chesonis and others left ACC and pondered their futures. This was right after the Telecom Act of 1996 was passed, and the Act, for the first time in the history of the United States, opened up all telecommunications markets to competition by all. The very term, CLEC, was coined as a result of the Act. Chesonis and a group of other executives from ACC decided to take advantage of the Telecom Act and formed PAETEC in 1998. His boss at ACC, Richard Aab, founded US LEC based in North Carolina, and Steve Dubnik, also of ACC, founded Choice One Communications in Rochester, NY, and after merging with CTC Communications in 2006, it moved its headquarters to Burlington, Massachusetts. So 1998 was a period in time when talented, experienced, entrepreneurial individuals became available to form telecom companies in markets that heretofore had been closed to them. They would create companies that would be alternatives to incumbent local phone carriers. The times were also ripe for venture capitalist money and potentially irrational exuberance to help provide start-up funding, motivation, and possibility thinking.

Chesonis and his PAETEC cofounders sought out the very best talented and motivated people to get PAETEC off the ground. These individuals believed in him and willingly took a chance on this startup in order to both build PAETEC and potentially create wealth for themselves as "everyone gets stock options; everyone has a share of the company they work for."[126] In order to have this opportunity, though, they had to have character. "PAETEC looks for *character*. Work ethic. Team spirit. A willingness to admit mistakes and strive to improve. That's the core competitive advantage — people who *care*, who value human relationships above everything else, relationships in the workplace, with customers, in the community and at home."[127]

PAETEC's success model seems to be based on the following essential, fundamental, interdependent, and synergistic three categories:

The first category is to have a positive value system and philosophy that emphasizes "ambitions, passion, caring, and feelings, which reflect life, substance, and values through interactions with other people."[128] This value system includes putting people first as the only way to do business and hiring the right caring individuals with character as paramount to properly building the company. Then it includes creating an environment conducive to creativity and teamwork. PAETEC employees become part of the PAETEC family. They are also encouraged to commit to the overall betterment of their community via community involvement by positive support and also by having PAETEC and its management model the same. A check of the PAETEC Web site under "COMPANY INFO/community connections" demonstrates probably the finest example of both documenting and communicating the importance of community involvement and support recently observed. There are many, many more important particulars that go into this fundamental category of values and philosophy. Perhaps they can be best summarized by something that seems to transcend countries, societies, and religions, and which has many versions and is called the Golden Rule, a.k.a., the Ethic of Reciprocity: Do onto others as you would have them do onto you.

The second category is a sound business plan that emphasizes profitability and success is the model's next fundamental category for success. One of the most basic differentiators between PAETEC's business plan and that of several other CLECs from the very start was that PAETEC targeted only small-to-medium-sized business accounts and above. Those CLECs that sought residential accounts and very small business customers via total resale or the purchase and resale of UNEs from ILECs under the guidelines of the Telecom Act of 1996 were the first to fail as specifically discussed earlier in this text. PAETEC's business plan is to target small-medium, medium, and in some cases larger clients where there is more potential for a successful profit margin. Then, it makes the customer the center of their corporate strategy by providing superior customer service at value pricing. The company's goal is to earn each customer's business every day. PAETEC implements various customer user groups and advocacy functions including satisfaction surveys after service processes, installation, and billing. It conducts something called Net Promoter customer surveys twice per year by asking a representative sample of customers the following Net Promoter question: "How likely are you to refer PAETEC to friends and colleagues?" So far, PAETEC has continued to delight customers, as they've consistently exceeded their internal goals for Net Promoter. Perhaps part of the reason is because 20% of each employee's bonus is tied to these Net Promoter scores and because there is a dedicated PAETEC Net Promoter manager responsible for Net Promoter and for identifying opportunities for continuous improvement. This ultimately benefits both clients and employees alike.

PAETEC avoids flashy TV ads and expensive, broad-reaching marketing campaigns in favor of word-of-mouth recommendations from existing customers to broaden their business customer awareness. They address the market via three channels. First there is a highly trained professional sales force with more than 500 direct salespersons. Approximately 65% of company revenue flows through this channel. Another 20% of the revenue comes from more than 800 sales agents who receive compensation for selling PAETEC communications products and for providing customer referrals. Their wholesale division provides the remainder of the revenue. Less than two dozen sales professionals sell PAETEC solutions to other carriers and resellers who may offer PAETEC communications service under their own brand.

Integral to providing caring service to their clients are two special areas where PAETEC really shines. First is their PAETEC Online Customer Care Center, which allows customers to integrate their business with PAETEC's. Considering that communications has more and more become the arteries that carry a company's lifeblood of orders and information, what customer wouldn't want this. PAETEC Online is free to every PAETEC subscriber. It allows for secure information transfer, account management, and problem resolution from any Internet-connected computer. Capabilities include a versatile set of information tools to help customers analyze service usage patterns, spot trends before they become problems, and plan for growth and change. PAETEC Online users can review bills and payment history and even stay abreast of industry events. There are specialized tools and functions designed for PAETEC agents as well. This proactive partnership allows for simple, informative, powerful, and responsive communications and information transfer via a secure link to PAETEC without substantial ongoing operational costs, so it delights both clients and shareholders.

The other very special area is PAETEC's Network Operations Center, a.k.a. the NOC. Once a client has service installed and working, PAETEC's NOC customer service representatives provide quality customer care second to none. Rather than resorting to industry standard interactive voice response systems (IVRs), they respond to customers using a personal touch — live customer service representatives who are caring, knowledgeable, well trained, and professional. This "results in nearly 90% of incoming customer service calls being answered in twenty seconds or less and approximately 35% of billing and service issues are resolved during the first call. Agents also own the responsibility

PAETEC Network
Operations Center
(NOC)
Courtesy of
PAETEC

for following up with each customer on the issue resolution progress."[129] Since their core competitive advantage is people who care and recognize that the customer is the center of PAETEC's corporate strategy, post-sale customer service is extremely high. The finishing touch to this customer-centered approach appears to be top management. PAETEC enables and requires their corporate executives to be involved with and responsive to clients well beyond typical industry norms and even beyond contractual terms and conditions. Mistakes do happen, but when they do and clients feel the need to escalate an issue to higher management, executive response has proven to be both fast and meaningful in such a way as to delight the client. This is no small accomplishment when there is a mistake and this outcome is not atypical, but rather it is deliberate and observably repeatable.

Completing their business plan is a broad and deep set of products and services that more than meet the needs of their clients. In addition to meeting immediate needs, PAETEC's product set is able to transform telecommunication client platforms from that of the twentieth century to the new technologies of the twenty-first century. PAETEC offers all the traditional business services including traditional voice, toll-free service, Voice over IP, multiple access types, data services including high speed Internet and e-mail, Web hosting, managed router support and MPLS VPN, and much more. It's interesting that in addition to offering core products similar to the breadth and depth of their much larger ILEC competitors, PAETEC offers even more through their Advanced Solutions Group (ASG). ASG is composed of six subsidiaries of PAETEC and provide products and services that complement PAETEC's core data and voice offerings and provide critical solutions for customers. After PAETEC purchased Allworx in 2008 to provide twenty-first century key and PBX premise-based equipment with computer telephony and networking integration capabilities, Allworx became part of the Integrated Solutions Group (ISG) ASG subsidiary, which also represents Avaya and Cisco and "handles installs and network management for complete VAR end-to-end connectivity on WANs and LANs."[130] PAETEC's Equipment for Services (EFS) program also is part of this entity and provides creative financing and subsidizes a customer's cost for new equipment or software as a part of a broader network services package. It may be possible to reduce or even eliminate the cost of upgrading their communications network.[131] PINNACLE Software (http://pinnsoft.com/) is another ASG subsidiary. It provides "a structured progressive approach to implementing service lifecycle management, consolidating disparate operations and data into a comprehensive whole, providing value-added self-service to their customers, and automating the provisioning of PBX and carrier services. They accomplish this through the PINNACLE Communications Management Suite. PINNACLE's product is compliant with key IP telephony solutions from Avaya."[132] It's rather unusual for a telecommunications provider to offer consultative energy services, but PAETEC offers this through yet another ASG subsidiary. Their Advance Energy Solutions subsidiary is somewhat unique to the industry. It differentiates itself in the marketplace by researching individual client energy profiles and then customizing a managed solution to meet client needs and satisfy budgetary requirements. Working together, ASG components coupled with PAETEC proper form a whole greater than the sum of its parts, thereby enabling greater 'through-sales' for PAETEC and greater 'through-solutions' for clients.

PAETEC's third and last fundamental category for success is to have a strategic, modern, cost effective, and efficient technology plan that is designed to be "evergreen" and potentially future proof. This concept and strategy allows PAETEC to be able to provide its services at a value cost to clients. It doesn't use thousands of miles of old facilities and hundreds of central offices that are thirty or more years old and housed in old buildings that require excessive HVAC. Rather, their state-of-the-art network is designed to be modern and efficient and now "green" concepts as well as sustainability principles are beginning to influence the PAETEC network. The core technology products that compose their network elements are provided by technology partners Cisco Systems Inc., Alcatel-Lucent, and others. Running a telecommunications company requires technology well beyond the network itself and PAETEC has built their IT/Operational Support Systems by using providers Dell; EMC; HP; Microsoft; Oracle; Sun Microsystems, Inc., and Harris Corporation. Intelligent Network design and optimization using top-shelf products from tier-one suppliers, coupled with operational efficiencies, help to reduce PAETEC's cost to provide service. This technology plan offers a synergistic path and management dashboard tool that enables its business plan to succeed.

While PAETEC's success model is formidable and the aforementioned categories of success and the actions within provided a sound base for success, they were not sufficient to drive PAETEC to its present status. It took something more. The element that has driven PAETEC to its present prominence as the largest CLEC in the United States is the vision that Mr. Chesonis and his team demonstrated a few years before when they created a longer-term strategic plan and then acted upon it when the time was right. They defined three very important issues PAETEC needed to address if PAETEC were to survive long term in the U.S. telecom industry. Mr. Chesonis said,

> The *first* area we had to focus on was becoming larger. At that point, we were a $600-million-a-year privately held company in twenty-eight markets, primarily in the Northeast. To compete with AT&T and Verizon for medium and large business customers, you had to have the size and the scale to make people feel comfortable that they weren't risking their job by becoming a customer or risking their family's security by becoming an employee. First we wanted to become a Fortune 500 company and we thought that a size of $4 to $5 billion in revenue would

help people feel comfortable that we had staying power. The *second* thing we had to focus on was that we needed national presence. You can't really compete for medium and large business customers unless you're in most of the markets. So our goal at that point was to be in 85 of the top 100 metropolitan areas in the country. The *third* issue was to get more of the fiber and wireless capacity from our data centers and major switching sites to our end users' buildings with facilities that we own or control. At that point, most of how we were connecting to people was through the incumbent local carrier or through an alternate provider. Being in larger tier-1 or tier-2 cities, we could still do that efficiently, but long term we still wanted about 15% to 20% of connections with customers to be our own fiber or our own wireless network. We always believed that by definition you need to have a majority of traffic and your business coming through connections that you own or control. Even after the last thirty years with all the different building that's gone on throughout the Internet boom and bust in all these cycles, over 90% of the buildings in the United States only have the incumbent local (telephone) company with actual facilities into the building, so you have to have a good resale model if you want to be successful. [133]

If we fast forward to the middle of 2008, we see that PAETEC acquired/merged with the North Carolina CLEC US LEC Corp. This acquisition provided the Southeast footprint and was with a complementary company very similar to PAETEC in terms of values, business plan, and technology. US LEC was founded by Mr. Chesonis' former boss at ACC, Richard Aab, who has now become a member of the PAETEC board of directors. This merger also allowed PAETEC to finally become a publicly traded company. It had been no small hurdle for PAETEC to go public, as they failed to do so during their first try in 2000 when the NASDAQ went from 5000 to 2000 points yielding a very hostile investment marketplace for an IPO. Then PAETEC failed to go public again in 2005 due once again to a hostile environment. Some financial consultants in the 2007 timeframe considered the movement to a publicly held NASDAQ traded company through the US LEC merger to be truly masterful. While the approach to "going public" was not traditional, it succeeded and allowed employees to finally realize their PAETEC stock options. Also in 2007, PAETEC announced a merger with McLeod USA Inc. Mr. Chesonis said: "The McLeod merger gave us the rest of the country and gave us some fiber facilities on top of that. They spent over $2.5 billion to install (fiber facilities) during their fourteen year run. After they went through two bankruptcies, we picked it up for a fraction of the $2.5 billion they spent."[134] PAETEC has not met all of its strategic goals, but is now well on its way. They're in nearly eighty-two of the top one hundred markets in the country and are around Fortune 1000 status, so they have both scope and scale. They also own a considerably greater amount of network facilities if not the 15% to 20% that they seek. Looking to the future, in a 2008 Rochester Business Journal interview, Mr. Chesonis said that future growth would come both organically through operations via gaining a larger slice of the market as well as through acquisition. "PAETEC couldn't grow organically fast enough to hit Fortune 500 status in a five-to seven-year timeframe."[135]

What, if any, are PAETEC's weaknesses, concerns, or unaddressed issues?

Clearly, PAETEC lacks a cellular wireless strategy. This is a conscious decision by PAETEC's corporate executives as wireless is clearly not a core competency of the organization. However, wireless is beginning to displace some traditional wireline access in the small-medium and above business market. So, too, are firms starting to desire to combine wireline and wireless billing when it makes sense. Clients sometimes also perceive that a unified wireless and wireline offer might provide more total value than two individual and separate offers. One example is the fact that discounts are often based upon total purchasing commitments. By adding together the wireless and wireline commitment that a corporation signs with a telecommunications provider, greater discounts might be appropriate. Another example might be integrating wireless and wireline voice mail functions via integrated mailboxes. The first example is strictly a financial benefit, while the second improves operational efficiencies. Currently, PAETEC does not address this evolving dynamic.

The other issue is one of concern and is best viewed from two vantage points that are opposite in nature. The issue is managing mergers. PAETEC has signaled the world that its intent is to not only grow organically, where they have demonstrated an excellent track record, but to also grow via mergers and acquisitions where they haven't yet proven themselves. From the viewpoint of a conservative outsider consultant or investor, there is uncertainty as a result of this statement. It is uncertain as to how effective PAETEC will be in its current merger endeavor, as PAETEC is new to large mergers and has not totally integrated them yet. The only thing that the stock market dislikes more than bad surprises is uncertainty. Today, it's still a *wild card* as to how successful PAETEC will be in this endeavor, so it's problematic. Until PAETEC creates and discloses a track record regarding the US LEC and McLeod USA mergers, this concern will exist due to uncertainty. However, from the vantage point of a current shareholder or investor or interested observer who believes in PAETEC's long-term strategic plan and greatly desires that PAETEC possess the economies of scale and mass that would strategically benefit PAETEC in the long run, there likely is the view that entering into this uncertain merger arena is not only positive, but highly desirable and even necessary. The reality is that PAETEC is definitely addressing this merger issue and is well on the way to completely integrating US LEC into the fold including the back-office changes. Likely by the end of 2010, even the complicated billing function will be complete. Then PAETEC will work on the McLeod billing integration. It is without a doubt that PAETEC understands the importance of this situation and places it as a top corporate priority. Likely, public communication relative to the success of the integration of mergers will play an important role in PAETEC's stock price in 2009 and 2010. PAETEC might be wise to demonstrate success and eliminate uncertainty via totally integrating the current mergers and acquisitions, optimizing operational efficiencies, and leveraging revenue growth before biting off more through the merger and acquisition route. Success in this area that is appropriately communicated to the public tends to emphatically eliminate uncertainty.

Let us close this PAETEC case study with an indication of the physical destination of the PAETEC headquarters, which also demonstrates innovation, creativity, and caring for the City of Rochester. In 1962, the first urban indoor mall in the United States named Midtown Plaza opened in Rochester, NY. It was tremendously successful for many years. This was a period of time when Rochester was very vibrant and employers like Eastman Kodak and Xerox enjoyed many times the number of employees at their Rochester headquarters buildings and locations than today. Over time, the predominance of the major employers and their work force declined and Midtown Plaza began a downward spiral. Time took its toll on the physical facilities and it lost its two main anchor stores in the mid 1990s and Wegmans Markets closed its grocery store there as well. Such situations were sometimes considered a sign of new urbanism, as such occurrences were often thought to be symbolic of suburban sprawl. It continued its struggles until October 16, 2007, with the announcement that Midtown Plaza would be knocked down via eminent domain with funding from the State of New York and the cooperation of the City of Rochester. A major part of the announcement indicated that replacing substantial portions of the plaza will be the new headquarters for PAETEC. This PAETEC headquarters building is being designed to adequately provide facilities sufficient to house all PAETEC headquarters employees likely to be needed in the future. PAETEC prefers to have all of its employees under the same roof as this allows for direct communication between employees and departments. This helps sustain its culture and also its positive, caring relationships with clients. The vision to relocate PAETEC headquarters from the suburbs to downtown is one more example of PAETEC caring about more than just PAETEC. The positive aspects of tearing down the aged Midtown Plaza, constructing modern office buildings, and providing over 1000 jobs in the area will greatly help to revitalize the City of Rochester. In a June 5, 2008 news release, Rochester, NY Mayor Robert J. Duffy said:

> Thanks to Governor (of New York) Paterson and Arunas Cheso-
> nis (PAETEC CEO), we have a long-term strategy to revitalize
> the heart of the city. That strategy involves capitalizing on the

city's unique assets and utilizing public-private partnerships to spur investment and greater economic development. This project will lay the foundation for recreating downtown as the place to live, work, and do business.[136]

Duffy went on to say:

The decisions that were made to create the Mall in 1958 are relatively the same ones that are facing us today. We are witnessing the same spirit to redefine downtown and create a vision for the city center with a community focus that Midtown architect Victor Gruen did in 1964 when he declared Midtown Plaza as the first living example of a theory of revitalization he called 'transfiguration.' Gruen felt that Midtown would provide a 'change of urban pattern, a new order.' This is our opportunity to be consistent with history and once again transfigure downtown.[137]

The new PAETEC headquarters is scheduled to open in 2013.

Case 1 Study Questions

1. Read the book: *IT ISN'T JUST BUSINESS; It's Personal* by Arunas Chesonis and provide a short report that shares three things that you learned from reading it.

2. Comment on the Golden Rule.

3. This case described three fundamental categories for PAETEC's success. Consider each and write a convincing argument regarding which of the three categories may have been most important to this success.

4. In your *opinion*, discuss the most obvious reason why those CLECs that failed actually failed. Provide a convincing argument that supports your opinion.

5. Explain what the term "evergreen" technology plan means and provide an example.

6. The term "sustainability" seems to be a key concept of 2008, 2009 and beyond yet networks were required to be sustainable long before this time period. How might communications companies build sustainability into their networks today?

7. Research and explain exactly why Mr. Arunas Chesonis and PAETEC stated what they did relative to the three important strategic plan issues that yielded the mergers and acquisitions of 2007 and 2008.

8. After reading this case study and performing your own research, put yourself in the role of a telecommunications industry consultant and write a one page executive summary that accurately describes PAETEC to your own clients.

9. Research and discuss the differences between a pure CLEC and that of an ILEC that creates a CLEC subsidiary. Why would an ILEC create a CLEC subsidiary?

10. Postulate the affects that PAETEC has had on the telecom industry.

Notes to Case Study 1

120 Arunas A. Chesonis and David Dorsey, *It Isn't Just Business, It's Personal*, Rochester, NY: RIT Cary Graphic Arts Press, 2006.

121 PAETEC, "Company Profile," http://www.paetec.com/strategic/PAETEC_profile.html.

122 "Shock and Awe," *Democrat and Chronicle*, 2009.

123 Connie Guglielmo, "Jobs to Get Treatment, Will Stay on as CEO," *Democrat and Chronicle*, January 6, 2009.

124 Chesonis and Dorsey, *It Isn't Just Business, It's Personal*.

125 Matthew Daneman, "PAETEC Raises Its Stock at Home," *Democrat and Chronicle*, July 9, 2007.

126 Chesonis and Dorsey, *It Isn't Just Business, It's Personal*.

127 Ibid.

128 Ibid.

129 Michele Pelino, "Case Study: PAETEC's Customer-Focused Strategy Captures Us SMBS," eds. Heidi Lo and Ellen Daley (Cambridge, MA: Forrester Research, 2008).

130 Richard "Zippy" Grigonis, "Allworx — Past, Present and Future," *Internet Tele-phony*, 2008.

131 Ibid.

132 Ibid.

133 "A CEO Who's Not Afraid to Think Big," *Rochester Business Journal*, 2008.

134 Ibid.

135 Ibid.

136 Robert J. Duffy, "Midtown Plaza Redevelopment Progress Announced," ed. Empire State Development, June 5, 2008.

137 Ibid.

Independent Telephone Companies
Embarq and Ontario Trumansburg Telephone

The Issue

Large players that were heretofore considered part of the telephone industry or the cable TV industry are converging applications and markets and morphing into a new and significant market sector that we will call Entertainment and Communications Providers. Verizon has its FiOS plan and AT&T has its fiber plan and there are other initiatives by leading former cable companies like Comcast Cable, Time Warner, and Cox Communications. They all are actively seeking that time in place when really large amounts of bandwidth are provided to end users via fiber to the home. The bandwidth available via Dense Wavelength Division Multiplexing (DWDM) and other advanced means over fiber optic cable is thought to be theoretically infinite. The resulting impact of all this available bandwidth will be vast and "game changing." Significant new and dramatic applications are forthcoming to those homes fortunate enough to reside in such geographic territories and there is a very large potential revenue stream that is incenting providers to make the colossal FTTH investment. Yet, there are still many "Never Bell" Independent Telephone Companies and likely many small cable companies that cannot make such large investments. They continue to deliver basic traditional or VoIP telephone service, Internet access, and/or cable TV via traditional means and may not be morphing into Entertainment and Communications Providers. Will these companies survive? What is their plan? How will residences within their territories address their digital futures? The purpose of this case study is to examine two such entities in order to enhance our understanding of the issue.

EMBARQ

When cell phone giants Sprint and Nextel merged in 2005, they spun off the local telephone business to be better able to focus on the jazzy new wireless company Sprint-Nextel. Thus, Embarq was born. At the time, most thought that Sprint-Nextel would be a superstar company, but the company has struggled with a host of merger-related problems and their stock has not performed well. Few expected much of Embarq because it focuses on the dowdy and some would say dying local basic telephone market, but so far its stock price has outperformed its parent Sprint-Nextel. Embarq is not really all that small. At its birth on June 1, 2006, it became the fourth- ranked telecom company in the United States behind AT&T, Verizon, and Qwest. It serves areas in eighteen states including Florida, North Carolina, and Nevada. There are approximately 18,000 employees and 6.8 million phone lines. Embarq is clearly a sizable corporation. What, then, is the problem? The bottom line is that the size of the basic telephone market is dwindling. "The local phone business dates to horse-and buggy days and is shrinking rapidly, thanks to alternative technologies such as the Internet and wireless phones. The number of U.S. local phone lines is shrinking about 6% annually — nearly 10% in some markets. Embarq isn't immune. The carrier is losing about 6% of its lines each year and revenue is down year-over-year."[138] Customers have many more options today than in the past. For instance, they may consider VoIP via cable providers, CLECs, or even nonfacility-based providers like Skype. Many, too, decide to make wireless their only voice communications option. At present, wireless-only households are in the 15% range and are primarily the choice for teenagers, young adults, and students in the 14-to-25-year old segments. Given this situation, let us now examine just how Embarq is addressing it.

In December 2004, Dan Hesse was named Chairman and CEO of the new company. Hesse had been head of AT&T Wireless before it merged with Cingular. "Hesse brought the leadership the company needed at that time as it literally embarked on its journey to be aggressive and innovative in combating two imposing threats: entrance into the wireline communications arena by cable companies and the growth of wireless-only households."[139] When Hesse moved on in December of 2007 (to become CEO of troubled Sprint-Nextel), Tom Gerke took the reins based on his experiences as a seasoned executive and with an eye toward the future by being aggressive, innovative, and feisty through developing products and services that allow customers to stay connected and get more done at home, at work, and on the go. Gerke had been Embarq's General Legal Counsel. Clearly, Embarq's number one strategy is to have leadership and vision as paramount characteristics in their CEOs.

This strong leadership has translated into a strategic and tactical emphasis on innovation, simplicity, and an enhanced customer experience so as to add value to Embarq's offerings and to optimize the user's experience via enhanced functionality, while not overcomplicating life. Rather than fighting the inevitable use of mobile devices by their customers, Embarq has turned this threat into a positive benefit. They've added features that integrate wireless and wireline via a "Mobile Virtual Network Operator" (MVNO) relationship with Sprint-Nextel that enables it to resell wireless service on the Sprint-Nextel network under the Embarq name. Hesse said, "We would not be in the wireless business just to sell wireless to customers. We are in the business because we think there is an untapped need to have wireless and wireline work together."[140] This concept is a primary differentiator for Embarq. He went on to say, "What makes us different (from other local phone companies) is our focus on innovation. That's not necessarily the first thing people think of when they think of a local phone company."[141] Embarq offers an integrated mailbox for wireless and landline phones. A missed call to either device triggers a message indicator on each. Therefore, users need only check one mailbox and are able to save time and effort. Another new feature allows customers to take wireless calls on their home landline and vice versa. Customers can have them ring simultaneously or select which phone rings first—wireless or landline—and in what sequence, for up to three phones. A wired/wireless integrated directory like your popular cell phone address book is under development. When a contact number is entered on a cell phone (or pc), it is stored on Embarq's network and can be accessed from cell and home phones. The integration of landline and cell phone is not just for consumers. They offer something called Embarq Smart Connect SM, which allows businesses to move calls seamlessly between their wireless and wireline networks without interrupting the call. Embarq also is the first company in the country to offer Embarq Smart Connect Plus, which allows calls to automatically move between Embarq wireless and on-premise Wi-

Fi networks, whichever is strongest. Clearly, the integration of wireless and wired telephone features and functionality is innovative and simple and enhances the customer experience by adding value. Embarq has made the value of wireless communications and wireline communications greater than each would have been by themselves in stand-alone modes. Embarq calls this integrated plan their "Together Plan." Undoubtedly, Embarq has high hopes that this added value will make a difference to customers' loyalty and that it will help to reduce the loss of their landline telephones and revenue stream.

What about the important "game changing" fiber-to-the-home strategy that the giants of the industry are pursing and betting their future on? It appears that over and above prudent fiber upgrades that are part of an evolutionary improvement of physical cable plant facilities, Embarq has no plans to commence a costly fiber-to-the-home investment. In two separate interviews, former CEO Dan Hesse states, "We believe we can grow the business without huge financial bets"[142] and in a bolder statement to Telephony Online, "We'd be better off putting that money in a savings account!"[143] He went on to say that for the foreseeable future also even in an improved business case for fiber; it would likely only apply to the densest third of Embarq's network. Hesse said, "Even if you go out ten or twenty years, two thirds of our video market is going to be satellite."[144] In a more recent interview, Harrison S. "Harry" Campbell, president of Consumer Markets for Embarq Corporation, said the following on this subject, "Video services complement our core voice and data services. We've taken a no-go decision on facilities-based video/IPTV service roll-out due to the expense involved. We are happy with our current partnership with EchoStar, which offers a great product and has strengths such as affordable packages, a strong high definition channel offering, and DVR (digital video recording). To us, it makes economic sense to continue to drive our EchoStar partnership instead of launching a facilities-based television service offering."[145] He went on to say that, "The demand for broadband or online video service is definitely exploding. These services offer the advantage of place-shifting and time-shifting, which appeals to a growing segment of consumers. We recently rolled out a so-called over-the-top or broadband video service through our portal, myembarq.com. The Embarq Video Store provides access to thousands of movies, TV shows, and music videos. In fact, this service gives our customers the opportunity to rent or buy/ burn content. These services are a good tool to increase ARPH (Average Revenue per Household) for broadband ISPs and are likely to play a more complementary role to conventional television services in the near future."[146] When asked what roles new and emerging services such as online music, gaming, video, home networking, and other ancillary services will have in Embarq's future consumer communications service offerings, Campbell said, "Emerging services will play a very important role from both customer retention and an incremental revenue perspective. Our focus is to keep increasing ARPH and once products like broadband access hit their saturation rate, other alternative products are needed. These auxiliary services will also enhance the overall customer experience."[147]

Rounding out an enhanced customer experience is the need to provide excellent service to customers and to the community at large. Embarq strives to surpass expectations in this area and has been recognized by the following:

- Ranking highest in customer satisfaction among large enterprise businesses, according to a J.D. Power and Associates study.
- Continuing the company's focus and commitment to the communities it serves. For example, in 2007 Embarq became only the third company in the history of the Greater Kansas City United Way to contribute more than $1 million.
- Embarq was named the 2007 Kansas City Business Ethics award winner for exemplifying high standards of ethical behavior.

Embarq is a relatively young company with a substantial heritage. It is small in size and in available investment capital compared to the goliaths of the industry, but still of substantial size when compared to other enterprises. In fact, Embarq is a U.S. Fortune 500 Company. It seems to understand the issue very well, and it also seems to clearly understand the colossal investment that facilities-based video/IPTV via FTTH would take. Evidently, its studies reveal that only in its major population areas will FTTH possibly make sense in the next ten years. Times change and with the changing times, outcomes of studies sometimes change as well. Embarq will undoubtedly continue to study facilities-based IPTV and other FTTH endeavors. In the meantime, Embarq's customers will have the opportunity to enjoy the added value that innovation and making communications services synergistic-but-simple-to-use affords. Those Embarq customers that fall outside of geographical territories where the emerging Entertainment and Communications Providers operate will not soon enjoy the extremely large access bandwidth and types of services that this will bring. Is this really that much different than a few decades ago (and in some geographic areas even today) when cable TV was *not* ubiquitously available in the United States? Just as "noncable-enabled" households were once prevalent, there will be a time and place where some consumers have gigabit and higher broadband to their homes and enjoy all the advanced products and services that FTTH will provide, while others won't. Perhaps this situation will be considered tomorrow's "digital divide."

Additional information regarding Embarq may be found at www. EMBARQ.com

The Finger Lakes region of Upstate New York State is often described as a beautiful and unspoiled area of forests, lakes, hills, villages, and vineyards. The region displays unique geological features that were created "thousands of years ago, when successive advances of the Ice Age continental glaciers thrust their fronts against escarpments extending across their path and into preglacial valleys."[148] This more than 9,000 square mile area also is one of the foremost tourist areas in the state and serves as the perfect backdrop for the largest wine-producing region east of the Mississippi. Residents, visitors, and tourists alike typically find this region appealing. Serving a portion of this area is a small, rural telephone company that turns out to be the largest family-owned independent telephone operating company in New York State. It's called

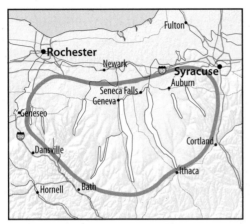

The Finger Lakes region of New York State

the Ontario & Trumansburg Telephone Company (OTTC). Like the Finger Lakes region itself, OTTC is unique and very interesting. Let us take this opportunity to learn about it and consider it relative to the stated issue of this case.[149]

According to a speech delivered to the Phelps Historical Society in 2003 by John Griswold,

> The first telephone service of any consequence in Phelps, New York prior to 1905 was established by Harry F. Flint, who formed a local company with the assistance of a Mr. Hubbell of the old Federal Telephone Company of Buffalo, NY. This was known as the 'Phelps Home Telephone Company.' In 1910, Mr. Flint sold the property to New York Telephone and remained in its employ until sometime later when differences arose. He left their employ and again through the assistance of Mr. Hubbell, formed the Phelps Mutual Telephone Company. New York Telephone had their exchange located over the Reynolds Hardware store and the Phelps Mutual exchange was located over the Bank. Because of his one dollar rate, Mr. Flint soon had most of his old subscribers back in his new company. The two companies continued in competition until March 12, 1920 when the New York Telephone Company again bought his company. Harry Flint was also the early provider of tele-

phone service in Clifton Springs until he sold it to the Federal Telephone Company of Buffalo. In 1917, the NY Telephone Company bought the Clifton Springs exchange from the Federal Company and operated it until June 1, 1920." [150]

Then, something occurred that most of us today find unusual and extraordinary. A forty year old named Hovey H. Griswold was a NY Telephone Division Plant Clerk in Syracuse, New York who had originally started with NY Telephone in 1903 sorting junk in Elmira, NY. Hovey Griswold was asked by his boss if he would like to own his own telephone company.

New York Telephone would give him their exchanges in Phelps and Clifton Springs if he would agree to assume the $18,000 mortgage outstanding on their telephone properties in the two villages. After several trips to Phelps and Clifton Springs and many sleepless nights, Mr. Griswold decided to accept the offer, give up his job with NY Telephone, and come to the area and form the Ontario Telephone Company, Inc. As of June 1, 1920, Mr. Griswold merged the two New York Telephone Company exchanges in Phelps and Clifton Springs into the new company with headquarters in Phelps. The new company had about 500 company-owned stations. The Clifton Springs Sanitarium owned and operated the Magneto Private Branch Exchange, but in March of 1926, Ontario Telephone Company purchased the PBX from the Sanitarium. The company installed a (then) modern common battery system that was large enough to provide not only for the business needs of the Sanitarium, but also telephones in the rooms of patients. [151]

The Trumansburg Home Telephone Company was purchased in 1926. Thus began what might be considered a little consolidation of providers in the area.

In the early fifties the decision was made to convert the company to dial telephone service. During the next four years, miles of lead cable were installed to provide facilities for improved service. A new general office building designed to be the headquarters of both the Ontario Telephone Company and the Trumansburg Home Telephone Company was built in Phelps and a new dial central office building was constructed in Clifton Springs. During the Open House to celebrate the completion of the new company headquarters in October 1954, Hurricane Hazel swept through the area uprooting trees and taking down many of the new cables that had been installed in preparation for the cutover to dial service. Despite this setback, the hard work of the company employees resulted in the successful conversion to dial service in December of 1955 for Phelps and in April of 1956 for Clifton Springs.[152]

Ontario and Trumansburg Telephone was one of the first in the state to introduce Direct Distance Dialing in 1962 and Touch Tone dialing in 1975. It's interesting that just after the turn of the millennium, the lead cable was replaced with fiber optic cable for feeder routes, and by 2003 OTTC had *"gotten the lead out"* of their territory and made available high speed Internet access and other broadband capabilities to over 90% of their customers. This capital investment was over $2 million. Note: In most cases, copper remained the primary mode of transmission for "last mile" residential services.

As a result of the United States introduction of competition to the telephone industry in the 1980s, OTTC formed a deregulated subsidiary called Comalert Systems Company, which allowed it to begin selling and servicing equipment outside of the OTTC-franchised areas. When the Internet exploded in the mid-90s, Comalert was reincorporated under the new name of Finger Lakes Technologies Group (FLTG) and it specialized in the integration of voice and data services including VoIP. FLTG is an integral part of the OTTC companies and is headquartered in Victor, NY.

With the courage of a true entrepreneur, Hovey H. Griswold gave up a good and safe position with New York Telephone in 1920, took a chance, and invested in what is now OTTC. He was joined by his son W. Malcolm Griswold in 1926 after he graduated from Syracuse University. Together they merged into the Trumansburg Telephone Company and greatly modernized operations. During the early 1970s Robert and John Griswold took over the reins of the company and continued to build on the progress of the past. In 1995 John's sons Bill and Paul took over the everyday operations of the company while the senior Griswold continued on the Board of Directors. Today, Hovey Griswold's great, great grandson Paul Griswold is president and CEO of the companies, now branded as the Ontario & Trumansburg Telephone Companies. He also heads Finger Lakes Technologies Group, Inc. The three companies are legally separate. The phone companies are regulated by the State Public Service Commission, while Finger Lakes Technologies, Inc. competes in its realm without telephone company regulation.

In 2008, OTTC is about a $19-million-a-year-and-growing enterprise. The rural phone company's base consists of 10,000 customer lines, providing slightly more than half of its revenues. While the revenues from this part of the business are holding steady or slightly growing, the quantity of lines is decreasing year over year but not to the 6% plus rate reported by Embarq. Virtually all of the revenue growth is from Finger Lakes Technologies Group. OTTC has fifty-one employees. Comparing OTTC to Embarq puts it in perspective. From a telephone line comparison, OTTC is about 0.2% the size of Embarq; from a revenue basis OTTC is about 0.3% the size of Embarq; and from an employee basis, OTTC is about 0.4% the size of Embarq. Yet it must address most of the same problems and issues of the much larger companies like AT&T, Verizon, Qwest, and Embarq. "The pint-sized company that Griswold has fashioned from the three firms is an ingenious and possibly unique amalgam," says Brighton, NY-based telecom consultant William Hughes, CEO of HPA Consulting."[153] OTTC's mission toward success, then, is to optimize the unique positive aspects of its situation and seize opportunity wherever it may be, while at the same time, diminish the outcome of any possible negative impacts that the environment and situation may have on OTTC. That seems to be exactly what OTTC President and CEO Paul Griswold's vision, direction, and action plan is accomplishing.

Let us examine this in more detail beginning with the 10,000 traditional residential telephone-line segment of the business. Just why is this quantity diminishing in numbers less than Embarq's and less than the overall U.S. average? Perhaps the marketplace itself has something to do with it. The Finger Lakes region is rural. There are not a lot of companies seeking out this area in order to expand their geographical territories and compete. In some areas, cellular coverage and service has only recently moved to good coverage and satisfactory levels. The area was not the first to enjoy cable TV service but now that coverage and service is good as well. It is safe to say that this rural market did not have as much competition in wired and wireless telecommunications or cable-TV as metropolitan and suburban markets. As a result, residential consumers did not have as many alternatives as consumers in the more populated areas. In addition, studies show that the greatest concentration of households that are wireless only is in the fourteen to twenty-five-year-old segment of the population. This is not a growing age group within the OTTC territory. Also, it is probable that there is less "churn" of people moving to and from this area as compared to metropolitan and suburban locations. Consequently, there is no natural redefinition of the type of telephone service that one chooses, so inertia sets in and consumers merely keep their existing service. These all are natural outcomes and conditions resulting from the market place. However, another reason that OTTC's basic landline telephone service is shrinking in quantity less than the national average may well be because of the excellent service and value that Ontario & Trumansburg Telephone Companies are providing their customers. OTTC treats customers not only like they wish to be treated but also like family. Some examples of this are:

- A visit to the OTTC web site (http://ottctel.com/index.php?option=home) demonstrates that the product mix for residential and business customers is up to date, attractive, and easy to order.
- OTTC was named Business of the Month by the Phelps Chamber of Commerce for January 2007 and 2007 Business of the Year by the Trumansburg Chamber of Commerce.
- In 1977 the employees of OTTC initiated a scholarship for graduating high school seniors within their territory. As we close out the first decade of the twenty-first century, funding from employees is matched with OTTC corporate funds and eight deserving high school graduates from high schools within the OTTC territory are awarded $650 via the Griswold Telephone Scholarship each year.

Let us now turn to the question of how OTTC is growing its revenue. First, let us indicate that President and CEO Paul Griswold has stated that OTTC has no plans in the near term to provide facilities-based TV and video services and that launching a capital construction campaign to install fiber to homes just would not have an acceptable return on investment at present. So for now and the foreseeable future, FTTH will neither provide revenue growth nor a large capital drain on investment, debt, and profitability. Yet fiber-optic deployment *is* at the foundation of revenue growth. In 2005, "Griswold strung a fiber network that initially connected the Ontario & Trumansburg companies' territories. Then, stringing cable to points where at least one significant business/ enterprise customer had signed up for service, he started to extend it. Cornell University was pleased to connect its Ithaca campus to facilities in Geneva and became the network's first tenant and main anchor. Ithaca-based Tompkins County Trust Co became another initial major tenant on the FLTG network. Over the past two years, Griswold has extended the fiber network to some 200 route miles with connections going through Canandaigua and Victor where FLTG has its headquarters and then to Rochester, New York. FLTG provides both lit and dark fiber. In dark-fiber deals, businesses lease space on a network but install and maintain their own optical electronics. In lit-fiber arrangements, the fiber company owns and maintains connections to the network and collects a fee for service."[154] During October 2008, Ithaca College hired Finger Lakes Technologies Group, Inc. to provide Internet bandwidth services under a five-year agreement. FLTG will provide redundant Internet bandwidth to the college through its fiber optic network that runs from Rochester, NY to Ithaca, NY.[155] FLTG is continuing to expand the reach and opportunities that providing low-cost fiber-optic links to businesses provides. It was a key player in the vision and planning of the $7.5 million, 180-mile fiber-optic system that is being deployed throughout Ontario County. The first phase, which includes a forty-mile stretch of fiber-optic cable from the Eastview Mall in Victor, NY to Cornell Agriculture and Food Technology Park in Geneva, was completed in August 2008. Ed Hemminger, the county's chief information officer and chief executive officer of the not-for-profit corporation set up to manage the project, said the design and engineering of the remaining 140 miles will start October 1, 2008, and the entire project will be finished by 2010. Hemminger said that "the project was established to provide high-speed connectivity for public safety, education, munici- palities, health-care, and prime business locations countywide."[156] Both Hemminger and Michael Manikowski, executive director of the Ontar- io County Industrial Development Agency, say that businesses have re- located to Ontario County, citing the fiber-optic ring as a key reason for their decision. An example of one such business is the Massachusetts- based technology firm Systems Maintenance Services, which opened a 3,500 square-foot center in Fishers, New York and is adding twenty jobs. Company executive Johnny Walker said that the ring was a draw. [157]

Another unusual and innovative opportunity is related to what was once the 10,000 acre Seneca Army Depot that now is in the final phases of decommissioning. Griswold agreed to lease a 750-acre sec- tion of this former Army depot and extended his fiber network through Seneca County to it. The site had been a main U.S. nuclear weapons storage site and is now owned by the Seneca County Industrial Devel- opment Agency. "Griswold's idea was to convert more than sixty above- ground bunkers into electronic record-storage sites and server farms. His initial plan was to spend some $7.5 million to prepare the bunkers and market the storage facility himself. But after running fiber-optic cable to the site and doing some cleanup of bunkers, he started negotia- tions with a firm that would set up and run the records storage business as a subtenant. FLTG will provide fiber connections and security and sub lease the space."[158] Clearly this is a unique, innovative, and gutsy move that fiber-optic cable deployment helped to facilitate along with Griswold's possibility thinking and vision.

The core mission of Finger Lakes Technologies Group started out to merely provide modern telephone systems to small- and mid-size business customers. This would include station key, PBXs, and now IP soft switches and often a conversion to lower cost and more modern IP telephony. While this is a contemporary opportunity, it is certainly not all that unique and innovative. It is, nevertheless, a positive grow- ing business that also allows FLTG to go beyond its initial mission by tapping into the synergy that exists by being a part of OTTC. It often orders additional business telephone lines from OTTC and, where possible, fiber connectivity when working with clients in either OTTC's geographic market or those that may be close to its now extended and diverse fiber network. This basic business opportunity, coupled with the synergy available by being under the OTTC umbrella, has provided revenue growth beyond what a typical stand-alone technology enterprise like FLTG would normally generate and that is great for the overall well being of the enterprise.

We've been able to uncover multiple positive and innovative examples of OTTC's recent revenue-generation accomplishments and their overall effect on OTTC seems to be very positive. It's also important to report that OTTC has not forgotten about the cost side of the ledger and the need for operational efficiencies. OTTC recently converted some of their old circuit-type central office switches to modern, low-cost, small-profile, energy-efficient IP switches. OTTC chose the VP3510 Class 4/5 softswitch from the leading provider of rural ILEC switches — MetaSwitch — to provide voice, long distance, unified messaging, and conference bridging. In addition to the efficiencies inherent in such a switch, this will enable OTTC to offer competitive VoIP telephone service outside of their traditional service area via the CLEC status that they set up in FLTG. Some months after their conversion, FLTG & OTTC President Paul Griswold said, "In order to remain competitive for another hundred years, we needed to expand and offer services outside of our traditional ILEC operation. MetaSwitch is a trusted partner as we build out this completely new system to offer advanced services to a new customer base."[159] OTTC was able to invest in their future without overleveraging their debt and thereby has been able to maintain a positive balance sheet. Prudent use of capital and a conservative approach to fiscal responsibility is a foundation of OTTC.

Providing fiber-based facilities to homes for video/IPTV and other advanced applications and the revenue that such services would generate is not a viable option for OTTC in the near future. However, having visionary leadership, coupled with both the ability to seize opportunities within and now, beyond their traditional marketplace, coupled with a conservative approach to fiscal responsibility, appears to be the major cornerstones of this company. Unlike Embarq, OTTC does not intend to offer advanced synergistic landline and wireless services at this juncture. Rather, OTTC will continue its current strategy of seizing opportunity wherever it exists and maintain operational efficiencies and fiscal responsibility. OTTC believes that this approach will continue to yield modest growth and success.

Additional information regarding Ontario Trumansburg Telephone Company may be found at: http://ottctel.com/

Post Case Actions and Ponderings

This case investigated one of the largest non- or never-Bell Independent Telephone Companies via studying the fourth ranked LEC in the nation, Embarq. The case also investigated one of the smaller non- or never-Bell Independent Telephone Companies by studying the family owned Ontario Trumansburg Telephone Company. A very interesting event occurred in late October 2008. Embarq announced that CenturyTel was purchasing it for a projected $11.6 billion. "If the deal is consummated, Embarq shareholders would own 66% of the combined company, while CenturyTel would own 34%. CenturyTel would also acquire $5.8 billion (part of the $11.8 billion) of Embarq's (relatively large) debt. Combined company senior leadership will be comprised of both CenturyTel and Embarq executives and corporate headquarters would be in Monroe, LA, with a continued presence in Overland Park, Kansas, which is the current home of Embarq."[160] Mike Sapien, a principal analyst for telecommunications consulting firm Ovum, wrote: "Without wireless or a major market density of customers, this combined entity only has a few places to go for innovation or growth."[161] He went on to question the merger and its possible success. In any case, based on Embarq's actions, it seems to imply that Embarq may have questioned its own abilities to generate success even though prior to this merger, they were/are the fourth largest LEC, and post merger they will be a larger company, but still ranked fourth. It's difficult to understand exactly how this growth and additional size will benefit the new company. It's interesting to note that, because of their current or future size, they will need far greater new and innovative revenue successes than a much smaller company like OTTC in order to have any profound effect on their bottom line. Having a large debt does not help their cause either. Further, Embarq and CenturyTel are publically traded corporations with shareholders who desire and, yes, demand growth and a respectable return on their investments, while OTTC is a private, family-owned company that may allow OTTC to better address hard economic times or periods of less growth should that develop.

Case 2 Study Questions

1. Research a small cable company and compare and contrast it to both Embarq and Ontario & Trumansburg Telephone Company.

2. Explain the so-called digital divide.

3. How does the "Case 2 issue" apply, if at all, to situations in countries other than the United States?

4. Research and update yourself on the current status of Embarq and/or the CenturyTel/Embarq merged company, as well as OTTC. Can you draw any conclusions from this research? Please explain.

5. Consider the residential customers within the OTTC geographical area. Comment on any concerns or insight relative to them.

6. Earlier it was indicated that technology, policy, market forces, and security all have major effects on the telecommunications industry. Research and consider future possibilities that may have either positive or negative effects on rural telephone companies or possibly their subscribers. From a telecom policy viewpoint, be sure to review the FCC Intercarrier Compensation proposal. From a technical viewpoint, research so-called "femtocells," a.k.a. "cellphone tower in your living room." Will they provide such good and enhanced fixed wireless service to homes that more residential customers will opt for wireless only?

7. How do telephone and cable companies compare? Select your favorite telephone and cable companies and write a brief "compare and contrast" paper.

Notes to Case Study 2

138 Leslie Cauley, "Spinoff Embarq Outshines Parent," USA Today, September 11, 2007.

139 EMBARQ, "Embarq History," http://www2.embarq.com/companyinfo/history/.

140 Cauley, "Spinoff Embarq Outshines Parent."

141 Ibid.

142 Ibid.

143 Ed Gibbons, "Embarq's Sophomore Season," Telephony, September 25, 2007, http://www2.embarq.com/about/articles/telephony_20070924.pdf.

144 Cauley, "Spinoff Embarq Outshines Parent."

145 "Movers & Shakers Interview with Harry Campbell, President Consumer Markets, Embarq," EMBARQ, http://www.frost.com/prod/servlet/exec-brief-movers-feature.pag?sid=124862916.

146 Ibid.

147 Ibid.

148 Oscar Diedrich von Engeln, The Finger Lakes Region: Its Origin and Nature, Ithaca, NY: Cornell University Press, 1988.

149 Emerson Klees, "Discover New York's Finger Lakes with Books by Emerson Klees," http://www.fingerlakes.com/.

150 Robert Griswold, "History of Ontario Telephone," Phelps, NY, 2003.

151 Ibid.

152 Ibid.

153 Will Astor, "Taking the Family Firm into a New Arena," Rochester, NY: Rochester Business Journal, February 8, 2008.

154 Ibid.

155 "Finger Lakes Tech. To Be College Web Provider," Democrat & Chronicle, October 27, 2008.

156 Julie Sherwood, "First Section of Fiber-Optic Ring Complete," Daily Messenger, August 23, 2008.

157 Ibid.

158 Astor, "Taking the Family Firm into a New Arena."

159 FLTG, "Seasoned Service Providers Learn New Tricks with MetaSwitch," http://ottctel.com/index.php?option=com_content&task=view&id=40&Itemid=93.

160 Allie Winter, "CenturyTel, Embarq Combo Lacks Wireless," RCR Wireless News, 2008.

161 Ibid.

Fiber Optics, the Non-carrier Opportunity!

United States and global citizens, businesses, applications innovators, users of high-speed IP links, and investors are all very much aware of the great demand for fiber-optic cable connections and the cable and electronics that help make up the wherewithal to allow these ultra broadband links to function. In 2007, Richard Mack of KMI Research/CRU, a London-based research firm, indicated that "Worldwide, we expect double-digit growth rates in fiber-optic cable demand for the next few years."[162] In China, fiber-optic cable demand has been in the 20% range for a few years, but may slow a bit as China has already built much of its city-to-city backbone. The 2008 summer Olympics in Beijing proved to be a stimulant as China proudly prepared to show off for the world. India too has a large demand appetite, mostly for Internet backbone expansion. Within the United States, the backbone carrier networks are pretty much complete relative to fiber. Unlike much of the rest of the world, the U.S. fiber-optic demand today is due in large part to the fiber-to-the-home Verizon initiative that projects a $23 billion budget through 2010. AT&T, while not as aggressive as Verizon relative to their initiative, has a large fiber plan that brings it to a node close to the home. Within the next ten years, the U.S. will likely become the first industrialized nation with tremendous broadband fiber-optic-based capabilities to the home. Verizon and AT&T and also some cable companies are making huge investments toward the day when they can provide vast amounts of broadband access to homes and deliver packages that might include multiple HDTV streams, video on demand, very high speed Internet and advanced gaming, and they may even consider including voice calling via VoIP as a "free" addition to such bundles since it takes little bandwidth compared to the other applications. Some consider it

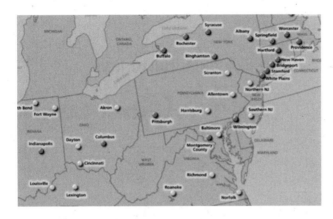

Fibertech 2008
footprint. Courtesy
of Fibertech

amazing and also gratifying that the large carriers now see tremendous potential revenue streams in the residential arena. It's amazing, because for some time this has been a rather depressed revenue-generating marketplace for them, so they have tended to favor and target the business market. Some may ask, what about Wi-Max? In specialized areas and also in many countries outside the United States, Wi-Max may be a nice broadband alternative to fiber for those areas that do not have the bandwidth needs that Verizon, AT&T, and cable companies are betting on in the United States. Wireless options like Wi-Max, while good, cannot compare to the bandwidth capabilities of fiber. This certainly benefits companies like Corning in Upstate New York, which is the prime supplier of much of the United States and worldwide fiber-optic cable, connectors, and some of the electronics.

Considering all this, it's no wonder that when we think of fiber-optic connections, we typically think of carrier-provided networks, links, and solutions. Yet with all the fiber advantages and opportunities, there is a noncarrier type of opportunity that this case study will reveal via introducing Fibertech Networks. Fibertech Networks LLC was formed in 2000, when there was a proliferation of start-up Competitive Local Exchange Companies (CLECs) resulting from the Telecom Act of 1996. The Telecom Act effectively broke the lock on the local exchange telephone marketplace by allowing new entrants to that heretofore closed-to-all-other-than-incumbent telephone companies. Basically, the Regional Bell Operating Companies and also the Independent Telephone Companies had a monopoly on telephone service within their geographic territories. The Act changed that to move to a more competitive marketplace. Many entrepreneurs entered this marketplace with great anticipation of riches to come. They remembered the opportunities and wealth that became available when the Bell System was broken up in 1984 and the lucrative long-distance arena was opened to competition. With the benefit of hindsight, it's unfortunate but true that the local telephone marketplace had much different economic factors than that of long-distance calling in the early 1980s. Nevertheless, these new CLECs charged into the local markets and developed a need for excellent metro-area; fiber-optic networks that could help them provide connectivity to their newly sought-after clients. Fibertech Networks was formed to help address this opportunity in 2000 and by 2005 had one of the nation's largest independently-owned metro-area, fiber-optic footprints, which included nineteen cities in and around the East Coast from Massachusetts to as far south as Maryland and from the coast to Indiana and Tennessee. It now includes twenty-one cities in its footprint.

This CLEC and carrier opportunity allowed Fibertech to become cash positive in 2002 and 2003 and to yield its first operating profit in 2004. Helping telephone companies, a.k.a. carriers, build out their networks was the opportunity that Fibertech initially seized, and, in spite of the 2001 crash of telecommunications in the United States, Fibertech was fortunate to experience a nice success curve from these opportunities. Many CLECs and others did not fare so well. Fibertech to this day continues to have a vibrant carrier solutions group and continues to offer both large and small carriers' metro-area, fiber-optic solutions to connectivity issues. It serves seven of the top eight U.S. long-distance providers and numerous regional CLECs, as well as local and wireless providers. However, beginning in 2002, Fibertech began shifting some of its strategic emphasis to enterprise clients, and this is the interesting opportunity that we'll discuss next.

Fibertech's sustained interest in the enterprise market was fostered by client and market demand. Early in 2000, Fibertech installed some fiber for enterprise clients, but due to the tremendous telecom carrier demand at the time, it declined further enterprise business, preferring to serve telecom carriers. However, after having presented at a Cisco Systems, Inc. conference in 2002, a line of people formed to talk with company officials about obtaining fiber. This was an under-served marketplace with tremendous needs. Fibertech began to take a fresh view of the enterprise business sector potential and began its long-term diversification toward a goal of an even split between carrier and enterprise customers and revenue. One might ask, Why the great demand for fiber in the enterprise world? The question brings a twinkle to one's eyes, as there seems to be an almost insatiable requirement for bandwidth. This seems like a new twist on the comments in the *Field of Dreams* story in which was said, "Build it and they will come." Without the benefit of fiber, most enterprises face bandwidth as a very restrictive, costly, and limiting issue for their applications and advances in informa-tion technology. However, once an enterprise decides to invest in the fiber solution, which yields sufficient bandwidth it owns, eliminating both financial and bandwidth limitations for future applications and progress is helpful. Basically, enterprises move from networks that are restrictive to an "enabling" network and mindset. Thus, the many and changing goals of the enterprise may be economically and technologi-cally "enabled," and the probability of enterprise-wide success and end-user satisfaction greatly increases. It's a winner!

Let me step away from Fibertech for a moment and outline what for me is one of the most significant, interesting, and now technically advanced applications that have been enabled by the use of fiber and the great and relatively inexpensive bandwidth that it brings to the table. Due to the obvious need for privacy, no client will be mentioned re-garding this application, although most enterprises have implemented some form of this application. The application goes by many names. I'll use redundancy, disaster recovery, and business continuance. In its most simple form, redundancy is merely providing for alternate paths of connectivity in case the main path has problems. At the link level, it's easy to see how fiber can greatly assist with this design. Adding an additional diversely-routed fiber link that is "up," but maintaining the level of throughput for the primary and additional link to no more than half of what would have been placed on a primary link alone allows one to move transmissions to a healthy link if one link develops difficulties. Stepping up from just the link level to a fiber ring that connects mul-

tiple essential devices, like mirrored database servers and processors, ex-
tends this back-up concept to the most essential end points in a network.
This helps maintain 100% availability and lets us all sleep more soundly
at night. Adding fiber connectivity and mirroring database servers and/
or processors, where complete enterprise processes may be housed,
helps to yield fast, efficient disaster recovery and enterprise-wide business continuance. In the twentieth century, the typical model was to
replicate data and, initially, physically send it (then it was bulk transmitted) to some off-site disaster recovery subcontractor where it would
reside, and, in most cases, nothing would be done with it. If a disaster
struck, personnel would relocate to the disaster recovery site and begin
processing with the hope that software and processing functions were
as up to date as the main processors. The twenty-first century makes
this model seemingly obsolete because fiber allows for on- or off-site
data mirroring and complete and synchronized databases available in
the event of a disaster. It also allows for completely updated processors
to be maintained exactly in synch with the main live processors with
the potential for actually using them for live production on a rotating
basis. This helps to eliminate many of the problems that were inherent with disaster recovery processes of the twentieth century. In effect,
fiber has changed the game relative to disaster recovery and business
continuance. Astute companies now not only buy into this concept but
consider it to be a basic cost of doing business in the twenty-first century.
Perhaps this change has occurred "just in time." Let us now leave this
twenty-first century, fiber-enabled and improved modern application
and get back to Fibertech.

So far, we've discussed the marketplace need or demand and the
great bandwidth capabilities available via fiber optics. While demand
and inherent fiber benefits got Fibertech off the starting line, their core
attributes of customer satisfaction, value, and dependability make Fibertech *the* company to deliver the goods in their service region. According to their Web site:

> **Customer Satisfaction:** With over a decade of network opera-
> tions behind us, we feel strongly that our outstanding customer
> satisfaction levels are a direct result of exhibiting unwavering

business integrity. We define this simply as consistently doing what we said we would do and then, when possible, going a step beyond. The most important secret to our success is found in providing a customer experience unlike any other. With things like your own dedicated, knowledgeable sales representative, record-time responsiveness, and bullet-proof product and service guarantees. Plus, every Fibertech customer has access to our technical support staff, 24/7. So if you ever need help, no matter what time of day or night, it's just a phone call away.[163]

Value and Dependability: We're committed to ensuring that the value of your relationship with Fibertech will be far richer than anything you'll ever pay for our services. It also includes our well-earned reputation for service dependency, low maintenance, and flexible solutions, in addition to such tangibles as outstanding customer satisfaction, unparallel quality, peace of mind, and loyalty. And while you may not be able to put a price tag on such intangibles, it's all part of the Fibertech package, at no extra charge.[164]

As carriers and enterprises continue to place a greater interest in redundant and diverse network connections, Fibertech Networks is helping to develop their IT networks with increased security and flexibility built in. Across twenty mid-size markets in the Northeast and Midwest, Fibertech is the alternative fiber infrastructure provider to the legacy telephone and cable companies. Leveraging this deep metro fiber optic network footprint, Fibertech offers highly-customized solutions to meet growing bandwidth requirement.[165]

Perhaps the best way to truly understand what the services of a company are capable of delivering is to review some examples. The following three Fibertech supplied Customer Case Studies will accomplish this for you. Last, and following these Customer Case Studies is a brief fact sheet snapshot of Fibertech, which includes its URL in case you wish to make a virtual visit to Fibertech.

One of the nicest parts of learning via this Case Study–Fiber Optics, the Noncarrier Opportunity — is that it's based on many innovations that fiber optic technology offers. The applications and customer case studies demonstrate real-world successes that have occurred already. My hope and view is that you, the reader, will see and experience even newer innovations in this area as time and change continues. It would be great if you would even create/invent some new and some improved applications yourself. Good luck!

CASEFILE »Wexford Capital LLC/Connecticut Fibertech networks

THERE'S NO HEDGING ON THE IMPORTANCE OF NETWORK PERFORMANCE

JOSEPH CURRAN, VP AND CHIEF TECHNOLOGY OFFICER AT WEXFORD CAPITAL LLC

❝ Thanks to Fibertech, we've future-proofed our business and put a system in place to maximize the benefits of our dark fiber network," says Curran. "Its high-performance, high-availability architecture provides the necessary bandwidth to support our growth. ❞

"Reduce risk. Increase reward." It's the first commandment for hedge fund companies and the key to performance—the critical factor in today's financial environment and how any hedge fund company is judged.

But achieving outstanding performance is no easy task in light of growing pressure to strengthen yields while overcoming challenging market conditions and increasing regulation. That's where the first commandment comes in for Joseph Curran, VP and Chief Technology Officer at Wexford Capital LLC, headquartered in Greenwich, Connecticut.

» Reduce Risk

Curran always has performance on his mind. With more than $5.5 billion of assets under its management, Wexford Capital cannot afford to take risks with its IT infrastructure.

"We're comprised of a diverse base of users—both highly technical staff and fairly demanding external investors, clients, prime brokers, and other business clients," says Curran. "We process vast amounts of time-critical applications, transactions and other information crucial to the overall success of our firm and clients."

But handling that amount of data is filled with risk. Interrupted service or lost data could be catastrophic in an industry where, perhaps more than any other, time equals money.

» Increase Reward

In order to gain a competitive edge, Curran is always exploring new processes and applications. It's his goal to mitigate risk while simultaneously making his systems more efficient and better able to respond to customers' wants and needs.

In order to do both, Wexford built an off-site data center to act as a virtual office in the case of interrupted network service. It also replicates mail archiving, journaling and every transaction that takes place within the company.

"At Wexford, we feel that the key to winning is aligning the technology department with our business objectives," says Curran. "Technology is a powerful tool that has leveled the playing field. Our focus is to center programs on sound, customer-focused IT investments that act as technology enablers.'

To meet the broadband requirements for this application, Curran turned to Fibertech for dark fiber connectivity between its headquarters location and data center facility. "Dark fiber provided us with unlimited bandwidth to connect our locations for the cost of about a T3 connection—a no-brainer as far as an operational expense," says Curran. "With this capacity, we can now treat a resource 50 miles away as if it was in the next rack over. We don't worry about quality of service, WAN compression, or optimization—our network is always on."

» Performance. How any company is judged.

Curran believes Web technology and security applications will continue to evolve and improve the way Wexford conducts its business. More and more, customers want instant access—anytime, anywhere—to critical information, and he believes Wexford can meet these requirements.

With the outstanding performance that Fibertech's network delivers to Wexford Capital, you might say that the future-proof is in the pudding.

»FIBERTECH IS A LEADING PROVIDER OF METRO-BASED FIBER OPTIC TRANSPORT SERVICES. VISIT **FIBERTECH.COM** TO SEE WHAT WE CAN DO FOR YOU.

CASEFILE »Monroe County #1 B.O.C.E.S. / ROCHESTER

SEEING THE POTENTIAL OF A SMARTER NETWORK

JOHN POLAND.//SUPERVISOR OF RESEARCH + DEVELOPMENT./MONROE COUNTY #1 B.O.C.E.S. »

Bandwidth is just not an issue. For a number of years, it was always an impediment to implementing the technology, and now it doesn't come up anymore.

As a cooperative of 19 school districts in Rochester, New York, the Board of Cooperative Educational Services (B.O.C.E.S.) serves a lot of very demanding, tech-savvy clients: nearly 100,000 students, faculty and administrators serving kindergarten through 12th grade.

'We sell services to schools, be it special education, busing for districts, just about any service they want to buy," explains John Poland, Supervisor of Research & Development for Monroe County #1 B.O.C.E.S. "The group that I'm involved with supports technology in the schools. We do everything from infrastructure to staffing to providing applications."

As with IT professionals everywhere, the last few years have seen John and his team devote more time and resources to building and improving the networks that connect their district clients' IT systems. "Over the years, we've built networks in each of those school districts. Our most recent project was to connect them all together back to B.O.C.E.S., providing them access to the Regional Information Center." And thanks to Fibertech, not just any access. After years of T1 connections and incremental upgrades, the districts will now be able to enjoy a state-of-the-art high-speed WAN Gigabit Ethernet.

»A WHOLE NEW CLASS OF OPPORTUNITIES.

The flexibility and virtually unlimited bandwidth of their new Fibertech network has allowed B.O.C.E.S. to take advantage of technologies like streaming video, says Poland. "We are up to about 600 videos that we are offering digitally. We're

previewing probably another 1,000. We're also looking at streaming real time." Rolling out this new service system-wide has only become a possibility with the help of the new network, he says. "We did some pilot projects initially, but to really do it, the direct fiber optic connection was going to be required."

Beyond the added benefit of a dramatic upgrade in the speed of their Internet access, Poland also foresees an increased use of advanced voice applications by the B.O.C.E.S. districts. In fact, it's already underway. "One of our district's is doing voice over IP across their network. I see us quickly expanding that to other districts."

»THE NETWORK VOTED MOST LIKELY TO SUCCEED.

Looking ahead, their new gigabit Ethernet network will also allow B.O.C.E.S. to connect to the ultra high-speed next generation Internet through NYSERNET, the New York State Education and Research Network.

It all means that whether it's today or well into the future, whatever applications and technologies B.O.C.E.S. wants to utilize to help their students excel, their Fibertech network will be ready. "Bandwidth is just not an issue," says Poland. "For a number of years it was always an impediment to implementing the technology, and now it doesn't come up anymore."

CASEFILE ·780·2··7935·   » Community Health Network / INDIANAPOLIS   fibertech networks

CUSTOMER POWERED NETWORKS

THE SEARCH FOR LONG-TERM NETWORK HEALTH

RICK COPPLE.//CHIEF TECHNOLOGY OFFICER./COMMUNITY HEALTH NETWORK »»

We're able to grow the network depending on our needs, on our timetable...

Based in Indianapolis, Community Health Network is an integrated health network that includes five tertiary-care hospitals, six immediate-care centers, three nursing homes, and a variety of other facilities. Their staff of 8,500 medical and support personnel handles approximately 37,000 inpatient admissions and 535,000 outpatient visits annually.

»NEW TECHNOLOGIES. NEW CHALLENGES.

The good news: new advances in information technology gave Community the opportunity to dramatically improve the quality and delivery of patient care. Digital medical libraries, advanced medical imaging such as Picture Archive and Communication Systems (PACS), and an all-digital "paperless" Heart Hospital were just some of the productivity tools being studied and implemented. The bad news? Those bandwidth-intensive applications, along with new federal data protection and disaster recovery requirements, meant that Community was quickly reaching the maximum capacity of its Wide Area Network (WAN).

»THE FIBERTECH DARK FIBER PRESCRIPTION.

Community was searching for a communications solution that could provide scalable bandwidth to support its growth, while at the same time significantly reducing costs. "It's the business case that drives our decisions, not technology," says Rick Copple, Chief Technology Officer for Community. "Without a business case and return on our investment, it won't get done."

After a nine-month evaluation of operational costs, anticipated technology needs, and telecommunications solution options,

Community chose Fibertech Networks. By connecting five of its locations with a fully diverse fiber optic ring, Community determined that it would see a total return on investment of more than $9 million over the term of its Fibertech contract.

Another key to Community's decision was Copple's belief that he could manage the network in-house, thus eliminating outsourced network service contracts. "We're able to grow the network depending on our needs, on our timetable, without having to change the entire topology," he says. "We have a more than capable staff to manage the network. In my opinion, my success, and my staff's success, is predicated on our ability to take advantage of opportunities like this to manage applications and infrastructure ourselves."

In addition to enjoying an exceptional ROI, Community is now well-positioned to support both its current and future growth. A dark fiber solution connecting its facilities provides it with unlimited bandwidth to support its IT initiatives, lower operational costs, and unprecedented network security and control.

Bottom line: Community has the ability to manage and support a robust network infrastructure that allows it to meet and exceed its administrative and patient care needs, while meeting its ROI requirements.

Fibertech Fact Sheet	
Year Founded	2000
Corporate Headquarters	http://www.fibertech.com/ 300 Meridian Centre Rochester, NY 14618 866-697-5100 (Phone) 585-442-8845 (Fax)
Ownership	Privately held
About Fibertech	As a premier broadband provider, Fibertech has quickly established itself as a high-quality provider of metro-based, fiber-optic transport services. Fibertech is a leader in building and operating fiber optic networks throughout mid-size cities in the Eastern and Central regions of the United States. The company has built metro-area networks strategically connecting local Telco central offices, carrier hotels, data centers, office parks, and other high traffic locations. In addition, Fibertech is unrivaled in its ability to extend its fiber networks cost-effectively into individual business locations to provide high performance, customized network solutions. Founded in 2000 and privately held, Fibertech investors are led by Nautic Partners of Providence, RI, and Bank of America Capital Investors of Charlotte, NC.
Markets Served	Fibertech operates one of the nation's largest, independently owned, metro-area fiber optic footprints in the United States. The company has core networks operational in twenty-one cities including Buffalo, Syracuse, Rochester, Binghamton, White Plains, and Albany, NY; Providence, RI; Pittsburgh, PA; Indianapolis, IN; Columbus, OH; Hartford, Stamford, Bridgeport, New Haven, New London, and Danbury, CT; Worcester and Springfield, MA; Concord, NH; Montgomery County, MD; and Wilmington, DE, with a lesser fiber presence in a number of other markets.
Leadership	Building on decades of telecom experience, Fibertech's management team has extensive expertise in all facets of fiber optic network design, construction, and management. Under the leadership of President and CEO John K. Purcell, with more than thirty years' experience with Frontier Corporation, and Frank Chiaino, with thirty years' experience in executive roles at Time Warner Cable, Fibertech is well positioned as a true business partner for its clients.
Customers Served	Fibertech serves major long-distance, CLEC, ISP, and wireless carriers. The company also boasts Fortune 500 enterprise companies, large financial institutions, major healthcare facilities, well-known universities and K-12 school districts, along with many mid-size companies in very diverse industries as customers.

Case 3 Study Questions

1. Consider the telecommunications environment in countries other than North America. Are there privately owned entrepreneurial fiber optic companies like Fibertech? If not, why not? Might this case offer an interesting concept and opportunity in other countries?

2. Where there are no such fiber oriented companies available, how is modern disaster recovery implemented?

Notes to Case Study 3

162 John Waggoner, "Investors Can Cash in as Demand for Fiber Optic Cable Takes Off," *U.S.A. Today* 2007.

163 Fibertech, "About Fibertech," http://www.fibertech.com/about.cfm.

164 Ibid.

165 Ibid.

166 ——, "Enterprise Solutions," http://www.fibertech.com/enterprise.cfm.

167 Ibid.

168 Ibid.

Glossary of Terms

Central Office: CO. (sē ō.) In the United States, this typically means a telephone company building where subscribers' lines are joined to switching equipment for connecting subscribers to each other and also to long distance providers. The European term would be a public exchange.

CLEC: Competitive Local Exchange Carrier. A CLEC is a new or competitive local phone company whose existence was made possible via the Telecommunications Act of 1996 which created the concept and term. The opposite of a CLEC would be an ILEC or Incumbent Local Exchange Company.

Collect Call: A telephone call in which the called person pays for the call.

Direct Distance Dialing: DDD is a telephone service that lets a user dial long-distance calls directly to telephones outside the user's local service area without operator assistance.

E.164: Is the ITU-T recommendation for GSTN (Global Switched Telephone Network) numbering. It is basically a sixteen-digit numbering standard that provides a unique telephone number for every subscriber in the world.

ENUM: ENUM is a proposal to map all phone numbers to IP addresses. The format for telephone numbers is specified in the ITU-T E.164 standard and the formats for URLs, IPv4, and IPv6 address are standardized by the IETF.

Equal Access: A term that came about at the time of divestiture. It provided that all long-distance common carriers must have Equal Access for their long-distance caller customers. City-by-city telephone subscribers were asked to choose their primary carrier, whom they would reach by dialing 1 before their long-distance number. All other carriers (including AT&T, if not chosen as primary) were reached by dialing a five-digit code (10XXX), thus providing Equal Access for all carriers.

FiOS: FiOS or Fiber Optic System refers to Verizon's proprietary Fiber to the premise service.

Horizontal Integration: A business strategy wherein an organization seeks to increase market share by selling the same or a similar product chain in multiple markets. Horizontal integration often occurs when a firm in the same industry and in the same stage of production is taken over or merges with another firm that is in the same industry and at the same stage of production (e.g., a car manufacturer merging with another car manufacturer). http://en.wikipedia.org/wiki/Horizontal_integration

IETF: Internet Engineering Task Force. Formed in 1986 when the Internet was evolving from a Defense Department experiment into an academic network, the IETF is one of two technical working bodies of the Internet Activities Board. Comprised entirely of volunteers, the IETF meets several times a year to set the technical standards that run the Internet.

Incumbent Local Exchange Carrier: ILEC. The Telecommunications Act of 1996 created the concept and term. Basically, an ILEC was an incumbent telephone company and had always had a protected geographic area within which, there was no competition for local telephone service. However the Act allowed for new or competitive local phone companies called CLECs. The opposite of an ILEC would be a CLEC. All RBOCs were ILECs but not all ILECs are RBOCs. ILECs are in competition with competitive local exchange carriers (CLECs).

Independent Telephone Company: A telephone company not affiliated with one of the "Bell" telephone companies. There were once over 1,400 independent telephone companies. The independent phone companies used to be represented by the United States Independent Telephone Association (USITA). Once divestiture happened, the association dropped the word "Independent" from its name, accepted membership of the Bell operating companies (but not AT&T) and the organization became USTA. Note: Independent Telephone Companies are ILECs, but they may also create a subsidiary to provide telephone services outside of their traditional geographical territory in order to provide competitive telephone services there. If they did this, they would also be considered a CLEC in those areas outside of their traditional geographic territory.

Intelligent Network: IN. This is an ITU-T concept designed to bring about a more efficient and easier method of providing advanced services to the public switched telephone network. IN and now, Advanced IN provide for such services as toll free service and the advanced features that are available within this service, custom calling features and number portability. SS7 in conjunction with software modules provide the intelligence.

Internet Protocol: IP. Part of the TCP/IP family of protocols describing software that tracks the Internet address of nodes, routes outgoing messages, and recognizes incoming messages. IP is used in gateways to connect networks at OSI network Level 3 and above.

Initial Public Offering (IPO): also referred to simply as a "public offering", is when a company issues common stock or shares to the public for the first time. They are often issued by smaller, younger companies seeking capital to expand, but can also be issued by large privately-owned companies looking to become publicly traded. http://en.wikipedia.org/wiki/Initial_public_offering

IPv4: Internet Protocol Version 4 is the current version of the Internet Protocol.

IPv6: Internet Protocol Version 6. This is the new Internet Protocol designed to replace and enhance the present protocol, which is called TCP/IP, or officially IPv4. IPv6 has 128-bit addressing, auto configuration and new security features and supports real-time communications and multicasting. IPv6 is described in RFC 1752.

Irrational Exuberance: This phrase was coined/first used at a speech to the American Enterprise Institute in 1996 by Alan Greenspan, then Chairman of the Federal Reserve Board. Shortly after his speech the stock market in Tokyo fell 3.2%, Hong Kong by 3%, Frankfurt and London by 4% and the Dow Jones followed suit and lost 145 points within the first thirty minutes of trading. Irrational Exuberance refers to the artificially overvalued state of markets, where investors and financial institutions are bullish and market sentiments are at a high. However, this bloated state is short lived and is shortly followed by panic and steep market slumps. Speech Excerpt: *But how do we know when irrational exuberance has unduly escalated asset values, which then become subject to unexpected and prolonged contractions as they have in Japan over the past decade?* http://en.wikipedia.org/wiki/Irrational_exuberance

IXC: IntereXchange Carrier. Also known as IEC (InterExchange Carrier) and IC. This was a carrier of long distance telephone calls but not local calls. The largest IXCs were AT&T, MCI, Sprint, and WorldCom. Note: This term is a bit passé since most major IXC's have been taken over by others.

Judge Harold Greene: Judge Greene presided over the 1982 AT&T Antitrust Settlement, enforcing its provisions and making decisions about requests from the participants to modify or reinterpret the provisions of the settlement. On August 5, 1983, Judge Harold H. Greene of the United States District Court for the District of Columbia gave final approval to a consent decree breaking up the largest corporation in the world — AT&T. The settlement, which concluded a 1974 antitrust suit filed by the U.S. DOJ to end the regulated monopoly that AT&T had exercised for decades over the U.S. telephone network, was attributed in large part to Greene's skills as a trial judge and to his gift for synthesizing highly complex issues. (http://www.ieee.org/portal/cms_docs_ip-ortals/iportals/aboutus/history_center/yurcik.pdf)

LATA: Local Access and Transport Area also called Service Areas by some telephone companies. This term was defined during the divestiture period in order to designate where RBOCs could offer service. They could offer local telephone exchange service within LATAs while AT&T could not. There were 196 local geographical areas in the United States within which a local telephone company may offer telecommunications services. At one stage, AT&T was expressly prohibited from offering intraLATA calls but the Telecom Act of 1996 makes it possible for all entities to offer local and long distance type of calling. Note: Since the Telecom Act of 1996 opened all portions of telecommunications to competition by all, this term is now somewhat passé.

LEC: Local Exchange Carrier. The local phone companies, which can be either a Bell Operating Company (BOC) or an independent (e.g., GTE and Frontier Telephone), which traditionally had the exclusive, franchised right and responsibility to provide local transmission and switching services within a geographic area.

Lightspeed: This is a proprietary offering from AT&T. Project Lightspeed was slated to employ FTTP (Fiber to The Premise) and FTTN (Fiber to The Node) and technologies to provide advanced telecommunication and entertainment services to 18 million homes by end of 2007. However, the target date has since been revised.

Modified Final Judgment: MFJ is the agreement reached on January 8, 1982, between the United States Department of Justice and AT&T and approved by the courts (Judge Harold Greene) on August 24, 1982 that settled the 1974 antitrust case of the United States versus AT&T. The MFJ divested AT&T of the local regulated exchange business and created seven regional holding companies–Ameritech, Bell Atlantic, Bell South, NYNEX, Pacific Telesis, Southwestern Bell, and US West. The MFJ placed restrictions on the RBOC local exchange carriers, namely that they could not offer long-distance communications.

NASDAQ: The NASDAQ (acronym of National Association of Securities Dealers Automated Quotations) is an American stock exchange. It is the largest electronic screen-based equity securities trading market in the United States. With approximately 3,200 companies, it has more trading volume per day than any other stock exchange in the world. It was founded in 1971 by the National Association of Securities Dealers (NASD), who divested themselves of it in a series of sales in 2000 and 2001. It is owned and operated by the NASDAQ OMX Group, the stock of which was listed on its own stock exchange in 2002, and is monitored by the Securities and Exchange Commission (SEC). http://en.wikipedia.org/wiki/NASDAQ

NANC: North American Numbering Council, pronounced "nancy." It is an industry council chartered by the FCC in October 1995 to assume administration of the NANP (North American Numbering Plan) from Bellcore, as well as to select LNP (Local Number Portability) administrators.

NANP: North American Numbering Plan. Invented in 1947 by AT&T and Bell Telephone Laboratories, the NANP assigns area codes and sets rules for calls to be routed across North America.

Natural Monopoly: This was a term that was often used during the period of time in the United States when one large telephone company controlled much of the nation. It was sometimes used by economists to justify regulation. The idea is that one company can provide certain services (such as gas, water, or telecommunications) considerably cheaper than two or three. Therefore, let one company have the monopoly on the service. The concept of economic regulation by the government is being replaced by the concept of free trade and competition.

Patent: A patent is a legal grant to inventors, limited in time and technological extent. It provides the right to exclude others from making, using, or selling the invention. The temporal extent of a U.S. patent is usually at least seventeen years.

Postalized Rates: A concept used to structure long distance telephone rates or prices so that they do not vary with the distance of the call but rather depend on other factors such as duration of a call. This is called "postalized" because it is analogous to putting a first class stamp on up to a 13 ounce properly sized envelope and having that be sufficient enough postage to mail from anywhere to anywhere in the United States.

Public Service Commission/PSC: The state regulatory authority responsible for communications regulation within a state. This is also known as Public Utility Commission PUC), Corporate Commission, and, in some states, the Railway Commission.

Public Switched Telephone Network/PSTN: This refers to the world-wide voice telephone network accessible to all those with telephones and access privileges. For the better part of the 20th Century, it was called the Bell System network or the AT&T long-distance network. Accompanying this concept was that of Plain Old Telephone service or POTs.

Quadruple Play: The provisioning of triple play services (High speed internet, voice calling, and television/video) with mobile service.

RBOC/Regional Bell Operating Company: This refers to the seven RBOCs created as a result of the Modified Final Judgment (MFJ). On January 8, 1982, AT&T signed a Consent Decree with the United States Department of Justice stipulating that at midnight December 31, 1983, AT&T would divest itself of its twenty-two wholly-owned telephone operating companies. According to the terms of this Divestiture Agreement, also known as the MFJ, those twenty-two operating Bell telephone companies would be formed into seven RBOCs of roughly equal size. The seven RBOCs (and the operating companies that formed them) were Ameritech (Illinois Bell, Indiana Bell, Michigan Bell, Ohio Bell, and Wisconsin Telephone), Bell Atlantic (Bell of Pennsylvania, Diamond State Telephone, The Chesapeake and Potomac Companies, and New Jersey Bell), BellSouth (South Central Bell and Southern Bell), NYNEX (New England Telephone and New York

Telephone), Pacific Telesis (Pacific Bell and Nevada Bell), Southwestern Bell (Southwestern Bell), and US West (Mountain Bell, Northwestern Bell, and Pacific Northwest Bell). (Note: In the term MFJ, it was the 1956 Consent Decree that was modified.)

Return on Investment (ROI): In finance, rate of return (ROR), return on investment (ROI) or sometimes just return, is the ratio of money gained or lost on an investment relative to the amount of money invested. (Wikipedia)

Semaphore: A system of visual signaling which typically uses flags or light positions. Semaphores were in common use for message signaling prior to the invention of the telegraph.

Signaling System 7/SS7: SS7 is an out of band signaling system that typically employs a dedicated 64-kilobit packet data circuit to carry machine-language signaling messages about each call connected between and among machines of a circuit switched network to achieve connection control. Signaling performs three basic functions: supervising, alerting, and addressing.

Telegraph: A system employing the interruption of, or change in, the polarity of DC current signaling to convey coded information. The telegraph was the first all digital telephone network.

Telephone: The telephone (from the Greek words tele = far and phone = voice) is a telecommunications device that is used to transmit and receive sound (most commonly speech), usually two people conversing but occasionally three or more. It is one of the most common household appliances in the world today. Most telephones operate through transmission of electric signals over a complex telephone network that allows almost any phone user to communicate with almost anyone. (http://en.wikipedia.org/wiki/Telephone)

TELRIC: Total Element Long Run Incremental Cost. TELRIC is a way of calculating the cost of phone service based on the incremental cost of new equipment and new labor, not counting the embedded cost of old equipment and the labor to install that old equipment. This concept, although likely initially unfair to ILECs, was introduced to create an easier path (resale) to local service competition than merely requiring new competing telephone companies to build and install their own network. Over time, this concept evolved to become more equitable to all.

Toll Call: A call to any location outside the local service area; a long-distance call.

Transmitter: The device in the telephone handset that converts speech into electrical impulses for transmission.

Triple Play: Triple Play services refer to the provisioning of voice, data, and video services offered as a bundled service and at a cost-effective rate. The philosophy underlying this marketing term is that a triple play environment will increase customer value while simultaneously generating more revenue for the service provider, thereby creating a win/win situation.

Unbundled Network Element: The Telecommunications Act of 1996 required that the ILECs (Incumbent Local Exchange Carriers) un-bundle their NEs (Network Elements) and make them available to the CLECs (Competitive LECs) on the basis of incremental cost as a method of allowing CLECs to more easily compete. UNEs are defined as physical and functional elements of the network, e.g., NIDs (Network Interface Devices), local loops, circuit-switching and switch ports, interoffice transmission facilities, signaling and call-related databases, OSSs (Operations Support Systems), operator services and directory assistance, and packet or data switching. When combined into a complete set in order to provide an end-to-end circuit, the UNEs constitute a UNE-P (UNE-Platform).

Universal Service: A system originally conceived by Theodore Vail to provide telecommunications service for everybody, including geographic areas that were difficult to access. This was achieved by slightly overcharging one set of customer's (municipal) to offset the high cost of laying telecommunication infrastructure for another group (rural customer). The goals of Universal Service, as mandated by the 1996 Telecom Act, are to:

- Promote the availability of quality services at just, reasonable, and affordable rates for all consumers.
- Increase nationwide access to advanced telecommunications services.
- Advance the availability of such services to all consumers, including those in low income, rural, insular, and high cost areas at rates that are reasonably comparable to those charged in urban areas.
- Increase access to telecommunications and advanced services in schools, libraries, and rural health care facilities.
- Provide equitable and nondiscriminatory contributions from all providers of telecommunications services to the fund supporting universal service programs. (www.fcc.gov/wcb/tapd/universal_service)

Vertical Integration: When a company expands its business into areas that are at different points of the same production path, for example, a car company that is expanding into tire manufacturing. A company such as this is often referred to as being vertically integrated. (http://dictionary.reference.com/browse/vertical%20integration)

Voice over IP/VoIP: VoIP (Voice over Internet Protocol) is a term that is generally used to describe telephone calls transmitted using the IP protocol. IP protocol does not use the typical circuit switched approach to call transmission but instead uses the address in the call to route through a packet network. The cost of VoIP calls is typically much less than calls transmitted over traditional circuit switched networks because they tend to use less bandwidth both due to compression as well as optimizing the available bandwidth. It is projected that eventually all telephone calls will use VoIP.

VPN/Virtual Private Network: A VPN is a network that uses the basic PSTN or IP network physical network as its base but then through software, segregates itself from the larger network thereby creating a virtual private network.

White Knight: Is a term used when a company or organization makes a more favorable offer to rescue an ailing company/organization when a third makes a bid for a hostile takeover of the ailing company/organization.

Wide Area Telephone Service/WATS: Is a term used to describe a discounted toll service provided by long-distance companies in the United States. It permits a customer to make calls to or from telephones in specific zones with a discounted monthly charge based upon call volume. AT&T created WATS. There are two types of WATS services: in and out WATS. In WATS was also referred to as 800-Service and is now called toll free service after the expansion to 8XX.

WiFi: Wireless Fidelity Wi-Fi is another name for a wireless local area network (LAN) running under the 802.11 a thorough z standards.

WiMax: Worldwide Interoperability for Microwave Access, WiMax is now also known as IEEE 802.16 and is a group of broadband wireless communications standards for metropolitan area networks (MANs) developed by a working group of the Institute of Electrical and Electronics Engineers (IEEE).

Note: *Newton's Telecom Dictionary* is the foremost reference resource for telecommunications, networking and information technology terms. It is highly recommended that all students of these disciplines purchase their own copy of this excellent resource for now and for future reference. If you would like a more detailed definition of these glossary terms or any other terms within these fields, please see *Newton's Telecom Dictionary*.

Bibliography

"A Brief History of Communications." Piscataway, NY: IEEE Communications Society on its 50th Anniversary, 2002.

"A CEO Who's Not Afraid to Think Big." *Rochester Business Journal,* July 25, 2008.

Ante, Spencer E. "Telecom Back from the Dead." *Business Week,* June 25, 2007.

Astor, Will. "Taking the Family Firm into a New Arena." Rochester, NY: *Rochester Business Journal,* Vol. 23, No. 26, February 8, 2008.

AT&T Archive. *Events in Telecommunications History.* Warren, NJ: AT&T, 1992.

AT&T. "Equal Access Ballot & AT&T Milestones." http://www.corp. att.com/history/milestone_1984b.html.

AT&T and McCaw. "People of AT&T Meet McCaw / People of McCaw Meet AT&T." Basking Ridge, NJ: AT&T, 1994.

Bastein, Donna, John Clark, and Geraldine Weber. "AIN: A Smarter Platform for Service." *Bellcore Exchange,* 1993.

BBC. "Microsoft Break-up Ruling Overturned." http://news.bbc. co.uk/1/hi/business/1361934.stm.

Boettinger, H. M. *The Telephone Book*. New York: Stearn, 1983.

Cauley, Leslie. "Spinoff Embarq Outshines Parent." *USA Today*, September 11, 2007.

———. "Verizon's Army Toils at Daunting Upgrade." *USA Today*, March 1, 2007.

Chesonis, Arunas A., and David Dorsey. *It Isn't Just Business, It's Personal*. Rochester: RIT Cary Graphic Arts Press, 2006.

Chodos, Alan. "Dick Tracy Watch." http://www.aps.org/publications/apsnews/199906/dicktracy.cfm.

Common Cause Education Fund. "The Fallout from the Telecommunications Act of 1996: Unintended Consequences and Lessons Learned," 2005.

Congress, U.S. "Telecommunications Act of 1996," 1996.

"Connecticut Research Report on Competitive Telecommunications." Dr. Richard G. Tomlinson, 1995.

Daneman, Matthew. "Five Cities Test High-Tech 911 System." *USA Today*, July 9, 2008.

———. "PAETEC Raises Its Stock at Home." *Democrat and Chronicle*, October 21, 2007.

Deak, James N. "North American Numbering Plan, Numbering Plan Area Codes." Bellcore Exchange, 1996.

Deloitte & Touche Consulting Group. "The Telecommunications Act of 1996—A Comprehensive Overview of the New Law." Atlanta: GA, 1996.

Duffy, Francis, Maureen Fiorelli, and Michael Wade. "Putting 800-Number Portability in Place." *Bellcore Exchange*, 1993.

Duffy, Robert J."Midtown Plaza Redevelopment Progress Announced." Edited by Empire State Development, June 5, 2008.

Dylan, Bob. "The Times They Are a-Changin." 1964.

EMBARQ, "Embarq History." http://centurytel.com/pages/aboutus/companyinformation/timeline/index.jsp.

Endlich, Lisa. *Optical Illusions: Lucent and the Crash of Telecom*. New York: Simon & Schuster, 2004.

ENUM. See Electric Information Center at http://epic.org/privacy/enum/.

Faulhaber, Gerald R. *Telecommunications in Turmoil*. Ballinger Publishing Co., 1987.

FCC. "Do Not Call Registry." http://www.fcc.gov/cgb/donotcall/.

———. "FCC Rule Making." http://www.fcc.gov/rules.html.

Fibertech. "About Fibertech." http://www.fibertech.com/about.cfm.

———. "Enterprise Solutions." http://www.fibertech.com/enterprise.cfm.

———. "Fact Sheet." 2007.

"Finger Lakes Tech. To Be College Web Provider." *Democrat & Chronicle*, October 27, 2008.

FLTG. "Seasoned Service Providers Learn New Tricks with MetaSwitch." http://ottctel.com/index.php?option=com_content&task=view&id=40&Itemid=93.

Freeman, Roger. *Telecommunications System Engineering*. New York: John Wiley & Son Inc., 1989.

Fulle, Ronald G. "Number Portability." University of Colorado, Boulder (1996): 132

Gallup Poll. "Gallup Poll National Number Portability Survey." 1994.

Gamble, Robert A. "Number Portability — What Happened." *Business Communications Review*, 1993.

German, Kent. "The Top 10 Dot-Com Flops." http://www.cnet.com/4520-11136_1-6278387-1.html.

Gibbons, Ed. "Embarq's Sophomore Season." *Telephony*, September 25, 2007. http://www2.embarq.com/about/articles/telephony_20070924.pdf.

Greene, Judge Harold H., ed. *United States V. American Telephone and Telegraph Company, 552 F. Supp 131*. D.C. Circuit Court, 1982.

Grigonis, Richard "Zippy". "Allworx — Past, Present and Future." *Internet Telephony*, (2008):38-49

Griswold, Robert. "History of Ontario Telephone." Phelps, NY, 2003.

Grosvenor, Edwin S., and Morgan Wesson. *Alexander Graham Bell: The Life and Times of the Man Who Invented the Telephone*. New York: Harry Abrams, 1997.

Guglielmo, Connie. "Jobs to Get Treatment, Will Stay on as CEO." *Democrat and Chronicle*, January 6, 2009.

Harrison, Crayton. "AT&T Disconnecting Pay Phone Business." *Democrat and Chronicle*, December 4, 2007.

Hatfield, Dale. "A Report on Technical and Operational Issues Impacting the Provision of Wireless Enhanced 911 Services." 2002.

Heller, Mike. "Phone Numbers on the Move." *Telephony* (1995): 46-50.

Howe, F. L. *Endless Voices — the Story of Rochester Tel*, Rochester Telephone Corporation, Rochester, NY, 1992.

Hundt, Reed E. "Words That Matter: Writing Down a Commitment to Competition and Kids Speech."(speech) April 2, 1996.

——. *You Say You Want a Revolution, A Story of Information Age Politics.* New Haven: Yale University Press, 2000.

Keating, Anne B., and Joseph Hargitai. "Ancient Networks." New York: New York University Press, http://www.nyupress.org/professor/webinteaching/history3.shtml.

Kennedy, Charles H. *An Introduction to U.S. Telecommunications Law.* 2nd ed. Artech, 2001.

Kharif, Olba. "Free Information Please." *Business Week*, April 23, 2007.

Klees, Emerson. "Discover New York's Finger Lakes with Books by Emerson Klees." http://www.fingerlakes.com/.

Lys, Thomas, and Linda Vincent. "An Analysis of Value Destruction in AT&T's Acquisition of NCR." *Journal of Financial Economics*, *v39* (1995): 353-78.

Maney, Kevin. "A Glance into the Crystal Ball Hints at a Future without 800-Numbers." *USA Today*, November 7, 2006.

Manning, Jason. "Ma Bell Breaks Up." http://eightiesclub.tripod.com/id310.htm.

Marquard, Odo. *Zukunft Braucht Herkunft*. Stuttgart, 2003.

McAleese, Robert J. "Conserving Area Codes." Bellcore Exchange, 1991.

Merriam-Webster Online Dictionary. http://www.merriam-webster. com/dictionary/

"Microsoft Unveils Unified Communications Product Road Map and Partner Ecosystem." http://www.microsoft.com/presspass/ press/2006/jun06/06-25UCGRoadmapPR.mspx.

Morgenstern, Gary. "800 Number White Paper from AT&T," AT&T News Release, 1992.

"Movers & Shakers Interview with Harry Campbell, President Con- sumer Markets, Embarq." EMBARQ, http://www.frost.com/ prod/servlet/exec-brief-movers-feature.pag?sid=124862916.

Ness, Susan. "End of the Beginning—Remarks by FCC Commis- sioner." http://www.fcc.gov/Speeches/Ness/spsn607.html.

Newton, Harry. *Newton's Telecom Dictionary*. 22nd ed. Berkeley: CMP Books, 2006.

PAETEC. "Company Profile." http://www.paetec.com/strategic/ PAETEC_profile.html.

Pelino, Michele. Edited by Heidi Lo and Ellen Daley. "Case Study: PAETEC's Customer-Focused Strategy Captures US SMBs.". Cambridge, MA: Forrester Research, 2008.

Ramsey Pricing Business Dictionary.com. WebFinance Inc., http:// www.businessdictionary.com/definition/Ramsey-pricing.html.

Robrock, Richard B. "AIN and Beyond — Putting a Vision to Work." Bellcore Exchange, 1995.

Sayre, Gregg. "Regulatory Distortions of Local Exchange Telecom-
    munications Infrastructure." Rochester Institute of Technology,
    B.Thomas Golisano College of Computing and Information
    Sciences, (2001):79.

Sherwood, Julie. "First Section of Fiber-Optic Ring Complete." *Daily
    Messenger*, August 23, 2008.

"Shock and Awe." *Democrat and Chronicle*, January 6, 2009.

Sorkin, Andrew Ross. "Charles Brown, 82, Former AT&T Chief, Dies."
    New York Times, http://query.nytimes.com/gst/fullpage.html?res=
    9F04E2D91738F930A25752C1A9659C8B63.

Spector, Robert. *Get Big Fast*, New York: HarperCollins Publishers,
    Inc., 2000.

Spekman, Robert and Bolon Meredith. "AT&T & Olivetti an Analysis
    of a Failed Strategic Alliance." (1993).

Standard and Poor's Industry Survey. (1994): 24-25.

Toth, Victor J. "Preparing for a New Universe of Toll-Free Numbers."
    Business Communications Review, 1995.

———. "Progress Report: 800 Database Access." *Business Communica-
    tions Review*, 1992.

von Engeln, Oscar Diedrich. *The Finger Lakes Region: Its Origin and
    Nature*. Ithaca: Cornell University Press, 1988.

Waggoner, John. "Investors Can Cash in as Demand for Fiber Optic
    Cable Takes Off." *USA Today*, September 24, 2007.

Weinman, Joe. "A New Approach to Search." *Business Communica-
    tions Review*, (October 2007): 19-29.

Wexler, Celia Viggo. "The Fallout from the Telecommunications Act of 1996." Common Cause Education Fund, 2005. http://hdl. handle.net/10207/4507.

Wikipedia. "Alexander G. Bell." Wikipedia, http://en.wikipedia.org/ wiki/Alexander_Graham_Bell.

——. "Bellcore." Wikipedia, http://en.wikipedia.org/wiki/Bellcore.

——. "Internet." Wikipedia, http://en.wikipedia.org/wiki/Internet.

——. "Jules Verne." Wikipedia, http://en.wikipedia.org/wiki/Jules_Verne.

——."Semaphore."Wikipedia,http://en.wikipedia.org/wiki/Semaphore.

——. "Semaphore Line Redirected from Optical Telegraph." Wikipedia, http://en.wikipedia.org/wiki/Optical_telegraph..

——. "Telephone Switchboard." Wikipedia, http://en.wikipedia.org/ wiki/Telephone_switchboard.

——. "Telegraphy." Wikipedia, http://en.wikipedia.org/wiki/Telegraph.

——. "The Thinker." Wikipedia, http://en.wikipedia.org/wiki/The_ Thinker.

——. "Triple Play (Telecommunications)." Wikipedia, http:// en.wikipedia.org/wiki/Triple_Play_Telecommunications.

Winter, Allie. "CenturyTel, Embarq Combo Lacks Wireless." *RCR Wireless News*, (2008): 1.

Yu, Roger. "GPS Becomes a Vital Tool for Frequent Travelers." *USA Today*, July 8, 2008.

Index

**And slowly answere'd Arthur from the barge:
"The old order changeth, yielding place to new..."**

—Alfred Lord Tennyson
Idylls of the King
"The Passing of Arthur"

 **green press**
INITIATIVE

RIT Press is committed to preserving ancient forests and natural resources. We elected to print this title on 30% post consumer recycled paper, processed chlorine free. As a result, for this printing, we have saved:

3 Trees (40' tall and 6-8" diameter)
1,475 Gallons of Wastewater
1 million BTU's of Total Energy
90 Pounds of Solid Waste
306 Pounds of Greenhouse Gases

RIT Press made this paper choice because our printer, Thomson-Shore, Inc., is a member of Green Press Initiative, a nonprofit program dedicated to supporting authors, publishers, and suppliers in their efforts to reduce their use of fiber obtained from endangered forests.

For more information, visit www.greenpressinitiative.org

Environmental impact estimates were made using the Environmental Defense Paper Calculator. For more information visit: www.papercalculator.org.